INTERNATIONAL POLITICAL ECONOMY SERIES

General Editor: Timothy M. Shaw, Professor of Political Science and International Development Studies, and Director of the Centre for Foreign Policy Studies, Dalhousie University, Nova Scotia

Recent titles include:

Pradeep Agrawal, Subir V. Gokarn, Veena Mishra, Kirit S. Parikh and Kunal Sen
ECONOMIC RESTRUCTURING IN EAST ASIA AND INDIA: Perspectives on Policy Reform

Solon L. Barraclough and Krishna B. Ghimire
FORESTS AND LIVELIHOODS: The Social Dynamics of Deforestation in Developing Countries

Kathleen Barry (*editor*)
VIETNAM'S WOMEN IN TRANSITION

Jorge Rodríguez Beruff and Humberto García Muñíz (*editors*)
SECURITY PROBLEMS AND POLICIES IN THE POST-COLD WAR CARIBBEAN

Ruud Buitelaar and Pitou van Dijck (*editors*)
LATIN AMERICA'S INSERTION IN THE WORLD ECONOMY: Towards Systemic Competitiveness in Small Economies

William D. Coleman
FINANCIAL SERVICES, GLOBALIZATION AND DOMESTIC POLICY CHANGE: A Comparison of North America and the European Union

Paul Cook and Frederick Nixson (*editors*)
THE MOVE TO THE MARKET? Trade and Industry Policy Reform in Transitional Economies

Mark E. Denham and Mark Owen Lombardi (*editors*)
PERSPECTIVES ON THIRD-WORLD SOVEREIGNTY: The Postmodern Paradox

Frederic C. Deyo (*editor*)
COMPETITION, POWER AND INDUSTRIAL FLEXIBILITY: Social Reconstructions of the World Automobile Industry

John Healey and William Tordoff (*editors*)
VOTES AND BUDGETS: Comparative Studies in Accountable Governance in the South

Jacques Hersh and Johannes Dragsbaek Schmidt (*editors*)
THE AFTERMATH OF 'REAL EXISTING SOCIALISM' IN EASTERN EUROPE, VOLUME 1: Between Western Europe and East Asia

Noeleen Heyzer, James V. Riker and Antonio B. Quizon (*editors*)
GOVERNMENT-NGO RELATIONS IN ASIA: Prospects and Challenges for People-Centred Development

George Kent
CHILDREN IN THE INTERNATIONAL POLITICAL ECONOMY

David Kowalewski
GLOBAL ESTABLISHMENT: The Political Economy of North/Asian Networks

Richard G. Lipsey and Patricio Meller (*editors*)
WESTERN HEMISPHERE TRADE INTEGRATION: A Canadian–Latin American Dialogue

Laura Macdonald
SUPPORTING CIVIL SOCIETY: The Political Role of Non-Governmental Organizations in Central America

Stephen D. McDowell
GLOBALIZATION, LIBERALIZATION AND POLICY CHANGE: A Political Economy of India's Communications Sector

Gary McMahon (*editor*)
LESSONS IN ECONOMIC POLICY FOR EASTERN EUROPE FROM LATIN AMERICA

David B. Moore and Gerald J. Schmitz (*editors*)
DEBATING DEVELOPMENT DISCOURSE: Institutional and Popular Perspectives

Juan Antonio Morales and Gary McMahon (*editors*)
ECONOMIC POLICY AND THE TRANSITION TO DEMOCRACY: The Latin American Experience

Paul J. Nelson
THE WORLD BANK AND NON-GOVERNMENTAL ORGANIZATIONS: The Limits of Apolitical Development

Archibald R. M. Ritter and John M. Kirk (*editors*)
CUBA IN THE INTERNATIONAL SYSTEM: Normalization and Integration

John Sorenson (*editor*)
DISASTER AND DEVELOPMENT IN THE HORN OF AFRICA

Howard Stein (*editor*)
ASIAN INDUSTRIALIZATION AND AFRICA: Studies in Policy Alternatives to Structural Adjustment

Geoffrey R. D. Underhill (*editor*)
THE NEW WORLD ORDER IN INTERNATIONAL FINANCE

Sandra Whitworth
FEMINISM AND INTERNATIONAL RELATIONS

David Wurfel and Bruce Burton (*editors*)
SOUTHEAST ASIA IN THE NEW WORLD ORDER: The Political Economy of a Dynamic Region

The Global Political Economy of Communication

Hegemony, Telecommunication and the Information Economy

Edited by

Edward A. Comor

Assistant Professor
School of International Service
The American University
Washington, DC

Foreword by Craig N. Murphy

Published in Great Britain by
MACMILLAN PRESS LTD
Houndmills, Basingstoke, Hampshire RG21 6XS and London

Companies and representatives
throughout the world

A catalogue record for this book is available
from the British Library.

ISBN 978-0-333-66477-3 ISBN 978-1-349-24926-8 (eBook)
DOI 10.1007/978-1-349-24926-8

Published in the United States of America by
ST. MARTIN'S PRESS, INC.,
Scholarly and Reference Division,
175 Fifth Avenue,
New York, N.Y. 10010

ISBN 978-0-312-12094-8 (cloth)
ISBN 978-0-312-16287-0 (paperback)

Library of Congress Cataloging-in-Publication Data
The Library of Congress has cataloged the hardcover edition as follows
The Global political economy of communication : hegemony,
telecommunication, and the information economy / edited by Edward A. Comor ;
foreword by Craig N. Murphy.
p. cm. — (International political economy series)
Includes index.
ISBN 978-0-312-12094-8 (cloth)
1. Information technology. I. Comor, Edward A. II. Series.
HC79.155G58 1994
384—dc20 93–44095
CIP

First edition 1994
Reprinted (with alterations) 1996

10 9 8 7 6 5 4 3 2 1
05 04 03 02 01 00 99 98 97 96

Contents

PART III DEMOCRATIC OPTIONS

Foreword: Communication and International Political Economy

In the early 1970s, the breakdown of the gold-dollar exchange standard, the first oil crisis, and the hullabaloo about Third World proposals for a new international economic order led to the establishment of international political economy (IPE) as a distinct academic field. Today the focus of the field is on the problems arising from the economic and political 'globalisation' that has been encouraged by recent revolutions in telecommunications and computers, the technological developments that the UN likes to join under the rubric of 'telematics'. Yet, surprisingly few IPE scholars study communication itself,[1] and those who do are apt to treat it like any other sector of the world economy, and not as a fundamental factor in the current global social transformation. Moreover, many IPE scholars remain unaware of the relevant scholarly work in Communication Studies, even the work by authors like William H. Melody who share the institutionalist framework frequently employed in IPE.

Edward A. Comor has done IPE a service by bringing together articles by Comor and other Communications scholars with work by political scientists and economists working within IPE. The result is a critical analysis of the deeper political and economic consequences of – and opportunities opened by – the deployment of new communication technologies. The contributors uncover much more than the rise of new sectors of manufacturing and services: inherent in the new communications industries are new possibilities for macro social orders, and new micro techniques of power that already have come to link the dominant and the dominated.

Viewed in one light, what Comor and his contributors simply point to is the latest manifestation of a much older pattern. For a century and a half – at least since the perfection of the electrical telegraph – technology has promised the abolition of distance and the globalisation of everyday life. Twice before – in 1865 with the creation of the International Telegraph Union and in 1906 with the creation of the Radiotelegraph Union – international agreements to encourage and then to regulate new international communication technologies have marked the beginning of generation-long conflicts over the boundaries of new, larger (but certainly less-than-global) economic orders. In 1920, John Maynard Keynes looked back on the 'economic Eldorado' of Europe's prewar Interimperial Order (which took final form only in the 1880s) with the same nostalgia that affects many of today's IPE scholars when they think back on the heyday of the Free-World Order, the postwar growth years that were anticipated in the Radiotelegraph pact forty years before.[2]

Perhaps some future historian will look back on the creation of the International Telecommunication Satellite Organisation (Intelsat) in the mid-1960s not only as the beginning of a third generation-long conflict over the shape of the global economic order, but as the vanguard of the globalised, co-operative world order that actually may take shape in the 1990s. But, right now at least, the communication revolution of which Intelsat was a part looks more like a harbinger of world disorder and a real contributor to today's global uncertainties.

As Ian C. Parker argues in his contribution to this volume, when faced with such radical uncertainty, people naturally turn to myths, to incomplete models of how the world works and how it could work if we all agreed on one vision of the future we want to share. Earlier, in similar periods of uncertainty (from the outbreak of the Franco–Prussian War through the Long Depression of the late-nineteenth century and again throughout the world wars and the intervening Great Depression) one prominent myth was that of a *laissez-faire* world in which all the 'globalisation' that could be achieved first by the telegraph and then by the radio (and by related revolutions in transportation) would usher in an era of peace and the greatest possible prosperity for all. The three case studies at the centre of this book (by Comor, Stephen D. McDowell and Milton Mueller respectively) reveal that this myth is still very much alive, not only as a guide to the liberal-fundamentalist policy of the Reagan administration, but, perhaps much more significantly, as one (albeit imperfect) way for what used to be called 'modernising elites' in the formerly relatively-closed Third World superpowers, India and China, to understand their own interests and aspirations.

In part because most of the authors of this volume have been inspired by Robert W. Cox's nuanced critical and historical IPE, they reflect an uneasiness that I share about the possible triumph of this liberal-fundamentalist myth. After all, access to communication channels opened by the telematics revolution is far from equal. Moreover, as comparison with earlier telegraph and radiotelegraph revolutions makes clear, what is really fundamentally new about today's technologies of global communication may be limited to the almost immeasurably greater amounts of information that now can be transferred across huge distances at the speed of light and the vastly increased capacity to manipulate and use that information. This results, as Martin Hewson suggests, in a vastly-increased capacity for surveillance and control by the dominant over the dominated.

The dominated (and those who would lead them), in turn, have already proved to be capable and creative users of those new communications technologies to which they have access: the few lines in the international telephone networks available to the faxers who helped prolong China's democracy movement, the hard-to-restrict access to Western broadcasts that helped embolden the democracy movements in eastern Europe, and the tape players and cassettes that helped to spread the word of so many anti-government movements in Africa, Asia, and Latin America. Of course, as both Adam Jones and James D. Halloran help us understand, such tools are available to those who honour the whole range of myths about the next social order: even if the 'wired world' may undermine the Orwellian capacity for ever-more complete surveillance, it does not necessarily mean the triumph of mutually-respectful 'multivocality', a granting of real political equality.

Yet, the authors of this volume's own respect for such multivocality (a characteristic of the broader critical theory that they share) helps assure that their open-ended analysis of the global political economy of communication is richer and more insightful than similar studies informed by a more limited theoretical agenda. Consider, for example, the more limited insights of liberal rationalists who recognise the political self-interest of many governments in maintaining government monopolies over international communication links, but who do not take into account the various myths and aspirations – nationalist and internationalist, liberal and mercantilist, charitable and 'realistic' – that can lead governments either to embrace those narrow interests or to reject them.

In contrast, this book does not just leave us with expectations about how a restricted calculus of power and interest will play itself out in one economic sector marked by rapid technological change. It leaves us with questions about how the new possibilities for collective human action, both coercive and cooperative, opened by new technologies, can be used to expand the truly voluntary realms of international civil society, rather than to consolidate the supremacy of the already powerful.

CRAIG N. MURPHY

Notes

1. Important exceptions include Jonathan Aronson and Peter F. Cowhey, for example their *When Countries Talk: International Trade in Telecommunications Services* (Cambridge, Mass.: Ballinger for the American Enterprise Institute, 1988).
2. John Maynard Keynes, *The Economic Consequences of the Peace* (London: Macmillan, 1971). Craig N. Murphy, *Global Governance and Industrial Change* (Cambridge and New York: Polity Press and Oxford University Press, 1994) discusses these earlier world orders and the role of transformations in communication technology in their origins.

Acknowledgements

Thank you to Emily Comor, Leo Panitch, Ian Parker, Steve Patten, Dino Rosati, J. Magnus Ryner, Tim Shaw, Tim Sinclair, Anne Stretch, Angie Schwartz, Graham Todd and Camus for their direct and indirect assistance in the preparation of this collection.

Finally, thank you to my parents for a lifetime of love and support. Without reservation, I dedicate this book to them.

EDWARD A. COMOR

List of Acronyms

ABC	American Broadcasting Company
AT&T/ATT	American Telephone and Telegraph Company
BCE	Bell Canada Enterprises
CBS	Columbia Broadcasting System
CD	Compact Disc recorder/player
CD–ROM	Compact Disc-Read Only Memory
CEO	Chief Executive Officer
CFC	Chlorofluorocarbons
CIS	Commonwealth of Independent States
CITIC–Pacific	China International Trust Investment Corporation
CNN	Cable News Network
COMECON	Council for Mutual Economic Assistance
CRTC	Canadian Radio-Television and Telecommunications Commission
CSL	Hong Kong Telecommunications Communications Services Ltd
CWHK	Cable and Wireless Hong Kong
DAT	Digital Audio Tape
DBS	Direct Broadcast Satellite
E-mail	Electronic mail
EC	European Community
EEC	European Economic Community
FAX/fax	Telefacsimile
FCC	Federal Communications Commission
FM	Frequency Modulation
FTA	Free Trade Agreement
GATT	General Agreement on Tariffs and Trade
GDP	Gross Domestic Product
GNS	Group of Negotiations on Services
GSM	General System for Mobile Communication
HDTV	High Definition Television
HKT	Hong Kong Telecommunications Ltd
HKTA	Hong Kong Telecommunications Association
HKTI	Hong Kong Telecommunications International
HKTUG	Hong Kong Telecommunications Users Group
HW	Hutchison Whampoa
IBM	International Business Machines

IIC	International Institute of Communication
IMF	International Monetary Fund
INTELSAT/Intelsat	International Telecommunication Satellite Organisation
IPE	International Political Economy
ITT	International Telephone and Telegraph Corporation
IVAN	International Value-Added Network
LP	Long-playing record
MFN	Most Favoured Nation
MODEM	Modulator–Demodulator
MPT	Ministry of Posts and Telecommunications
NATO	North Atlantic Treaty Organisation
NHK	Japanese Broadcasting Corporation
NIC	Newly Industrialising Country
OECD	Organisation for Economic Co-operation and Development
OPEC	Organisation of Petroleum Exporting Countries
PBX	Private Branch Exchange
PEC	Political Economy of Communication
PNTS	Private Non-Exclusive Telecommunications Service
PRC	People's Republic of China
PTT	Post Telephone and Telegraph
RTHK	Radio-Television Hong Kong
TELEX/telex	Automatic Teletypewriter Exchange Service
TNC	Transnational Corporation
TV	Television
UN	United Nations
UNCTAD	United Nations Conference on Trade and Development
UNESCO	United Nations Educational, Scientific and Cultural Organisation
VCR	Video Cassette Recorder
VHS	Video Home System
VSNL	Volume–Sequence–Number Line

Notes on the Contributors

Edward A. Comor is an Assistant Professor at the School of International Service, The American University, Washington, DC. His research interests include political economy, communication and culture policy, and the sociology of human knowledge. Professor Comor has conducted research for various organisations, including the Canadian Senate, and has published articles on the Canada–United States Free Trade Agreement, Canadian communication and culture policy, and the political-economic thought of Harold A. Innis.

James D. Halloran is the Director of the Centre for Mass Communication Research at the University of Leicester. From 1972 to 1990, he was the President of the International Association for Mass Communication Research. The author of numerous publications, Dr Halloran has received honorary doctorates from the Universities of Tampere (Finland) and Bergen (Norway). In 1991, he was the winner of the McLuhan Teleglobe Canada Award, the most prestigious award in international communication studies.

Martin Hewson teaches international relations at the Department of Political Science, York University, Toronto. He writes on multilateralism and the world information order.

Adam Jones is a PhD candidate in Political Science at the University of British Columbia and a lecturer in Latin American Studies at Vancouver Community College. His recent publicatons include a co-authored chapter in R. Brynen (ed.), *Echoes of the Intifada*, and articles in *New Political Science, Ethnic and Racial Studies*, and *Socialist Alternatives*.

Stephen D. McDowell lectures at Carleton University in Ottawa. His research interests include communication policies in Canada and India, international trade and investment in services, and the study of policy research programmes of international organisations.

William H. Melody is Director of the Centre for International Research on Communication and Information Technologies (CIRCIT) in Melbourne, Australia. He was Founding Director of the UK Programme on Information and Communication Technologies (PICT) in London, and has held positions at St. Anthony's College, Oxford; Simon Fraser University, Vancouver; and the University of Pennsylvania, Philadelphia. Among his publications on communication industries, economics and public policies, William Melody has authored five articles on telecommunication and computing issues in the *Canadian Encyclopedia.*

Milton Mueller teaches at the School of Communication, Information, and Library Studies at Rutgers University in New Jersey. He is the author of *International Telecommunications in Hong Kong: The Case for Liberalization* and *Telephone Companies in Paradise: A Case Study in Telecommunications Deregulation.* He is currently completing research on the history of universal telephone service in the United States and of trade and telecommunications in greater China.

Craig N. Murphy is Professor of Political Science at Wellesley College and Co-Director of the College's International Relations Programme. He has written widely on International Political Economy and North–South relations. His most recent publications include *Global Governance and Industrial Change* (Cambridge and New York: Polity Press and Oxford University Press), a history of the economic impact of international institutions since 1850, and 'International Institutions, Decolonisation, and Development', an article in the *International Political Science Review* written with Enrico Augelli. Both Professor Murphy and Ambassador Augelli are currently working on a new book entitled *International Leadership.*

Ian C. Parker is Associate Professor of Economics at the University of Toronto. He has published numerous articles on the economics of communication, Harold Innis, and trade and development, and has edited or co-edited several books in these fields. He is currently co-authoring a book on the treatment of information in economics.

1 Introduction: The Global Political Economy of Communication and IPE[1]

Edward A. Comor

The tendency of social scientists to work within their own specialised fields and the negative consequences of this tendency are much discussed but rarely redressed. As a result of the production of mountains of books and articles and the self-perpetuation of specialised academic languages, the time and effort required to resist specialisation has increased despite the widespread introduction of computerised research resources. As Robert Cox points out, specialisation provides 'a necessary and practical way of gaining understanding'.[2] Nevertheless, to deal with questions involving social change through space and time, 'a structured and dynamic view of larger wholes' remains an essential undertaking.[3]

Today, the characteristics of capitalism – globally, nationally and locally – demand a substantive advancement of our creative–intellectual capacities. Throughout the 'developed' world, for example, unprecedented structural change is taking place involving the ascendancy of service and information activities. This has both reflected and facilitated fundamental alterations in the global political economy which have themselves been stimulated by unprecedented transnational communication capacities. Assuming that the development of new international trade agreements involving the liberalisation of telecommunication services and a universally enforceable intellectual property rights regime continues, as do recent patterns of global transnational corporate (TNC) mergers and acquisitions, it appears likely that the technological capacity *truly* to transcend historical, temporal and spatial communication barriers will arrive by the beginning of the next century.

THE EMERGING INFORMATION ECONOMY

In 1986, the output of the international information economy – including mass media, electronic services, communication equipment and components – was worth $1.185 trillion (US), or almost 9 per cent of the world's recorded economic output. During the decade of the 1980s, the measured value of world telecommunication services grew by 800 per cent. In the United States, the largest producer of communication hardware and software, more than two-thirds of the work force are now engaged in jobs directly involving the production, processing or distribution of information.[4]

Through the convergence of computer and telecommunication technologies, with the application of fibre optic cables, satellites, signal-compression and digital technologies, an historic political, economic and social 'revolution' is supposedly under way. Unprecedented costs and stakes are involved. In the United States alone, estimates of the cost of constructing a fibre cable network linking every household range from $200 to $500 billion (US). In the construction of this so-called 'electronic highway', all forms of existing and not yet imagined electronic services will be integrated over what essentially will be a 'single wire'. Not surprisingly, these costs and stakes have led powerful vested interests to promote the 'common sense' notion that the emerging information economy is not only 'inevitable' but also 'good' for the world as a whole.

Largely due to the rapidity of these developments, and the role of these vested interests in 'just letting them happen', related implications and contradictions have barely been addressed in an analytically vigorous way. To paraphrase Thorstein Veblen and Harold Innis respectively, invention has become the mother of necessity, and it is the rapid development of these 'necessary' communication capacities that appears, paradoxically, to have undermined our capacity to understand.

It is this concern with 'understanding' that compels a careful review of what a political economy of communication (PEC) perspective may have to offer students interested in international political economy (IPE) and what IPE may have to offer PEC. In this introduction, I provide a general description of these perspectives and present some preliminary suggestions as to how they may be considered to be complementary. My focus will be on how PEC may be used to complement the development of IPE. I believe that readers more familiar with IPE than PEC will find much in this collection that may stimulate new directions in their work, and this introduction is largely addressed to them.[5]

THE POLITICAL ECONOMY OF COMMUNICATION

There are at least two extreme yet interrelated levels at which PEC can be used to complement IPE. For the sake of simplicity, these can be seen to occupy 'micro' and 'macro' levels of analysis. Unlike most IPE writers, many PEC scholars investigate 'the audience' and its relationship to information and entertainment producers and/or distributors and/or the means of distribution.[6] In most of these analyses, the audience is considered as a participant in what has been termed the ongoing 'social construction of reality'.[7] PEC scholars also recognise that the producer/distributor relationship with the audience involves both the explicit and implicit exercise of power. The participants in this power relationship possess and usually put to use their material resources and intellectual capacities.

A PEC perspective tends to focus attention on the material circumstances involved in the communication process. Importantly, this process, and the capacities of its participants, affect and are affected by the physical capacities of the communication media (or medium) used in the process. How and what human beings think (and therefore how they act) is directly conditioned by the communication process, who controls it, and how this control is exercised.

This concern with both the audience and the communication process – a concern with what I have termed the 'micro' level of analysis – *necessarily* impels a concern for large-scale global developments in communication (the 'macro' level).[8] Through the convergence of computer and telecommunication technologies, the virtually instantaneous transmission of database information, a software programme, or a televised soccer match simultaneously to billions of people is now *technologically* feasible through direct broadcast satellites (DBS) and the construction of transnational optical fibre-based networks. However, technological feasibility does not necessarily imply political, economic, or cultural feasibility. To some degree, in recognising these potential barriers, PEC scholars again tend to focus on the importance of power capacities, from the 'local' level through to the 'global'. This usually involves a study in the development and application of resources in the explicit and implicit construction and reproduction of what Antonio Gramsci called 'common sense'.[9] It is precisely this concern with the complex processes of globalisation and its manifestations that, at the macro level, directly links PEC to IPE.

INTERNATIONAL POLITICAL ECONOMY[10]

Robert Cox's work on the 'internationalising of the state', involving coalitions, alliances and historic blocs of social forces, constitutes the most promising analysis of communication in IPE. Both explicitly and implicitly, information and the production and development of social knowledge are fundamental to his work which has been directly influenced by Gramsci and his concept of 'hegemony'. As Cox explains,

> Hegemony is a structure of values and understandings about the nature of order that permeates a whole system of states and non-state entities. In a hegemonic order these values and understandings are relatively stable and unquestioned. They appear to most actors as the natural order. Such a structure of meanings is underpinned by a structure of power, in which most probably one state is dominant but that state's dominance is not sufficient to create hegemony. Hegemony derives from the dominant social strata of the dominant states in so far as these ways of doing and thinking have acquired the acquiescence of the dominant social strata of other states.[11]

Stephen Gill adds, 'Hegemony ... can never be the simple product of the predominance of a single state or grouping of states exerting power over other states ... partly because human beings have consciousness and a degree of free will or agency within the limits of the possible'.[12]

But while students of IPE have raised questions involving 'consciousness' and 'free will', PEC scholars have addressed these in a more vigorous and sustained fashion. I have mentioned the work of Cox and Gill because in many ways their writings constitute the most 'developed' of IPE theorisations on these issues. And while the space is not available here to present an overview of their work, a critical assessment of at least one key concept developed and used by Cox – his category of 'ideas' – will be used to illustrate how a PEC perspective may be used to complement IPE. First, however, a very general overview of what have been called the 'realist' and the 'complex interdependence' perspectives to IPE will be provided in order to contextualise Cox's self-proclaimed 'critical' approach. I do this in order to enrich the canvas on which we may evaluate PEC's potential contribution.

What have been termed realist approaches to IPE consider the nation-state as the basic unit of analysis. As a result, international relations reflect an inevitably conflictive (Hobbesian) struggle among nation-states. Realists therefore assume that (in the absence of a world government) inter-state relations are inherently conflictive; that the primary motivation underlying this conflict is the quest for power and security; and that human beings are naturally group oriented and that the nation represents an expression of this essence. As a result of these assumptions, the realist associates a stable world order with the hegemonic dominance of one or more nation-states. Unlike Cox's Gramscian use of hegemony, however, the realist understanding of hegemony involves little more than the presence of disproportionate military and economic capabilities. As with other realists, Robert Gilpin therefore considers that the decline of the United States post-1945 economic dominance (itself largely the result of rising foreign competition and the internationalisation of production activities) will result in the decline of international political and, ultimately, military stability.[13]

Another IPE approach, often referred to as the liberal view, which adopts what has been termed the complex interdependence perspective, assumes that the structures and competition characterising the global capitalist marketplace, rather than the nation-state, constitute the basic framework for analysis. Because private sector activities are fundamental, and because the world order is characterised by the complex interaction of these interests, nation-state bureaucracies, as well as diplomatic and military relationships, the term 'complex interdependence' has been adopted.[14] On the subject of hegemony, the complex interdependence perspective considers that the existence of a web of international economic, military and institutional interests presupposes and reinforces a collective interest among nation-states for maintaining the international status quo – especially among predominant countries and TNCs. Thus, in contrast to the realist, the complex-interdependence theorist assumes that the relative decline of a hegemon's economic and military capabilities does not necessarily imply a subsequent period of instability. Instead, the establishment of a complex of international regimes may induce a co-operative rather than a conflictive international environment.[15]

In sum, the realist perspective lacks a detailed recognition that world order and disorder involve much more than the relationships of state-based actors. The liberal complex interdependence perspective, while recognising the primacy of capitalist activities, fails adequately to address the role of civil society – that is, political parties, educational institutions, religious organisations, community groups, trades unions, the media, and many others. In this general absence, the liberal fails to conceptualise the presence or potential of a relationship among institutions and organisations with information, knowledge, or, using Cox's terminology, 'ideas'. In other words, neither realists nor liberals have displayed a recognition of the importance of the intellectual capacities of elites *and* masses and the role of these capacities in shaping state and private sector activities. As Robert Cox writes, 'The hegemonic concept of world order is founded not only upon the regulation of inter-state conflict but also upon a globally conceived civil society'.[16] Simply put, in an age of unprecedented international communication developments, realists and liberals lack a detailed understanding of 'the audience'.

ROBERT COX, 'IDEAS', AND HEGEMONY

In Robert Cox's 'critical' approach, what is called the state/society complex is considered to be the basic entity of international relations. Cox's Gramscian perspective hence enables the theorist to go beyond the narrow conceptualisation of international relations as state or regime relations. The importance of this integrated approach lies in its theoretical potential in analysing how the interests of elite groups are mediated with the thoughts and material circumstances of the masses. In Gramsci's words, 'Undoubtedly, the fact of hegemony presupposes that account be taken of the interests and tendencies of the groups over which hegemony is to be exercised, and that a certain compromise equilibrium should be formed'.[17] In sum, Gramsci conceptualises hegemony as more than just a complex form of domination. More precisely, the concept represents a *process* that involves the capacity of a class to engage in and dominate institutional developments and, when necessary, control the form of mediated compromises.

Cox conceptualises three reciprocal 'categories of forces' to interact within a given historical structure: *Material Capabilities, Institutions* and *Ideas.* Briefly defined, 'material capabilities' are the natural, technological and organisational resources available or potentially available (e.g. oil reserves, university research facilities). By 'institutions', Cox is referring to organised reflections of a particular order, including expressions of prevailing power relations (e.g. the international banking system, a political system, the labour union movement). Finally, 'ideas' entail shared practices and meanings as well as differing perspectives (e.g. diplomatic conduct, economic theory, gender roles).[18] Cox applies these forces in the form of a dialectical triad. This triad is used by Cox as an heuristic device and it is put to use when he applies 'structural' ideal types in his analyses.[19]

Another feature in Cox's methodology worth mentioning is his use of core–periphery analysis. According to Cox, the historical basis of international hegemonics has been the outward expansion of the national hegemony of a powerful state and historic bloc. In both *pax Britannica* and *pax Americana*, Cox uses the term hegemony in reference to more than just the dominance of a single power. Instead, hegemony involves the 'dominance of a particular kind where the dominant state creates an order based ideologically on a broad measure of consent'.[20] An incipient world order develops through interstate relations and the states themselves become more interlinked through the internationalisation of institutions and policies. Importantly, Cox's core–periphery model is not analytically wed to either diverse geographic locations or interstate relations *per se*. From the perspective of a transnational corporation, for example, the 'core' can constitute a group of full-time career employees engaged in senior-level managerial, financial and marketing activities, while the 'periphery' can include employees with less job security, ranging from middle managers down to largely unorganised unskilled workers seeking occasional contracts. The importance of including a non-geographic core–periphery model lies in Cox's point that the peripheralisation of workers into distinct groups itself undercuts labour's capacity for collective action.[21] And, as recent historical developments indicate, this enlarged core–periphery conceptualisation constitutes a relatively accurate reflection of contemporary post-Fordist patterns.

According to Cox, the decline of an explicit US hegemony and the emergence of today's neo-liberal order has involved the development of a complex form of international state–corporate interdependence. 'Countries', writes Cox, 'could not secede from a system to which they were bound in a web of reserve holdings, foreign indebtedness, foreign investments, trade outlets, and political and military commitments. The neo-liberal state had become a tributary to an uncontrolled world economy.'[22] During the 1970s, surplus production capacities led to an extension of Third World 'elite' consumer markets and, for some (mostly 'high-tech') industrial sectors, the massive overhead costs involved in research, development, retooling and marketing necessitated the globalisation of the production process (including markets).

Faced with a limited growth in consumer demand and mounting international debt, 'core' capitalist countries and transnational corporations faced three general options:

1. An intensification of inter-corporate (and inter-state) competition and/or symbiosis;
2. A radical redistribution of income and a shift away from elite-targeted production;
3. A revolutionary takeover of transnational production capacities and their redistribution on the basis of autonomous and/or collective state action.[23]

Because options 2 and 3 are fundamentally antithetical to the structural framework of international capitalism and against the interests of those directly affiliated with this framework, core states and transnational corporations have pursued the first option.

Cox understands the global reshaping of state structures to have been a result of a combination of both external pressures arising out of changes in the international structure and domestic social-economic realignments. Cox writes that advances in the internationalisation of the state, on the one hand 'provokes counter-tendencies sustained by domestic social groups that have been disadvantaged or excluded in the new domestic realignments'.[24] On the other hand, the internationalisation of the state itself is premised on (1) 'a process of interstate consensus formation regarding the needs or requirements of the world economy that takes place within a common ideological framework'; (2) a hierarchically structured basis for the formation of this consensus; and (3) adjustments in the internal structures of states in order to facilitate the transformation of global consensus into national policy and practice.[25]

From a PEC perspective, a number of critical points can be raised. Cox's internationalisation process, in which state tendencies are countered by domestic opposition, conveys an under-theorisation of *process*.[26] This under-theorisation is apparent in that Cox's internationalisation process lacks a specific theory of communication and, as a result, Cox is limited in his attempt to elaborate on the quantitative and qualitative characteristics of the internationalising state.

Cox conceptualises 'ideas' as both 'intersubjective meanings' (common norms of shared behaviour) and 'collective images' (various, often opposing, conceptualisations of the nature of prevailing power relations or 'ways of seeing'). For Cox, 'ideas' reflect both 'what is thought' and the existence of 'alternative thoughts' but no account is given as to the potentials and/or limitations of thought. As a result, 'ideas' are largely considered the results of an ongoing interaction with other elements. The great difficulty here is that we are left without a means of evaluating, for example, the influence of 'material capabilities' and 'institutions' on 'ideas'. As a result, our ability to recognise what *has* shaped 'ideas' may be attainable through a detailed reading of history. However, our capacity to project what *will* shape 'ideas' is limited.

In sum, Cox's conceptualisation of 'ideas' lacks a necessary theorisation of *the capacity for thought* in both the individual and the collectivity, and the dynamics conditioning its production and reproduction.

This criticism should not be read as some sort of curt dismissal of Cox's approach. On the contrary, the complexity, flexibility and scope of Cox's 'critical theory' means that it constitutes perhaps the most sophisticated and precise theorisation of international relations to date. In fact, Cox's description of what he calls 'critical theory' – of which he acknowledges to being a practitioner – contains one of the fundamental inspirations for this book:

> A principal objective of critical theory ... is to clarify [the] range of possible alternatives. Critical theory thus contains an element of utopianism in the sense that it can represent a coherent picture of an alternative order, but its utopianism is constrained by its comprehension of historical processes. It must reject improbable alternatives just as it rejects the permanency of the existing order. In this way critical theory can be a guide to strategic action for bringing about an alternative order.[27]

In step with Cox's emphasis on developing our 'comprehension of historical processes' in order to use critical theory as 'a guide to strategic action', the chapters that follow provide a basis on which to develop IPE through PEC and PEC through IPE. Before proceeding with an overview of these chapters, a brief example illustrating the limitations inherent in Cox's methodology may be instructive. In his widely read *Production, Power, and World Order*, Cox quotes Susan Strange at length in explaining the internationalisation of state policy. Strange considers this process to be the product of a systemic tension involving the national defense of sovereignty versus the necessity of complying with the demands of creditors and transnational corporations. '[O]ne certain result,' says Strange, is that both national officials and their international opposite numbers will have a 'strong incentive to fudge the issues and to conceal and obscure any possible conflict between the national interest of the debtor-states and the national (or special international) interests of the creditors.'[28] Strange also sees a tendency to mute differences as a reflection of the shared outlook of financial professionals (including economists) whose training and work-related socialisation provide 'no feasible alternative to the policies adopted'.[29]

While this tension certainly exists, it is unclear why Strange's 'fudgy' outcome has emerged without a serious and sustained counter-challenge. In other words, why has the conflict involving the defense of national sovereignty and the systemic interests of international capital resulted in the widespread adoption of a neo-liberal brand of 'common sense' among nation-state officials? Again, why are 'options' 2 and 3 mentioned above – a radical redistribution of income *or* a revolutionary takeover of transnational production capacities – so 'unrealistic'? Or, to put it another way, why has no 'realistic' alternative emerged from where Cox expects it to emerge – the social-economic periphery (i.e. the working class and/or the Third World and/or New Social Movements, etc.)?

Perhaps most fundamentally, Cox's explanation of these remains underdeveloped in the sense that his approach lacks a systematic conceptualisation of *how* institutions and other mediation nodes shape the capacity of human agents to conceptualise hegemonic alternatives.[30] Here, Cox emphasises the importance of the Gramscian concepts of an 'historic bloc' and the role of 'organic intellectuals'. Cox, like Gramsci, defines the latter as 'a corps of persons around whom a coalition of social forces ... [can] be constructed'.[31] It is this elite (usually sharing similar social-economic experiences) that holds the creative and structural capacity to define the 'national interest' and, in so doing, transcend and redefine economic interests, including the ability to convince weaker, more vulnerable sectors of the need to co-operate. At the apex of an emerging global class structure is a 'transnational managerial class' whose ideology, as expressed through the institutions that mediate it, penetrates into state and related institutions. As a result, senior personnel tend to adopt the view of the 'top' (or the 'centre') and adapt it to meet their particular conditions.[32] Given the emergence of international interdependence, and the rise of world competition, Cox considers the basis of organic intellectuals' shared 'worldview', and their capacity to convince others of its necessity, to lie largely in 'the risk of economic failure with a drastic drop in living standards' facing any country seeking some kind of defensive withdrawal from the world economic system.[33]

This is certainly a logical conclusion. But to sustain the dialectical vigour of Cox's own methodological framework, the question of specifying process again must be raised. Cox *begins* to develop this in his application of a complex core–periphery model. Because this model is not strictly based on geographic location, the peripheralisation of workers into distinct groups can itself be seen as the precondition undercutting workers' capacity for collective action and, hence, a counter-hegemonic consciousness. Fundamentally, this process directly involves communication capacity. Or, from the position of Cox's methodology, it is a question involving the nature of *how* 'ideas' are formulated and *how* they interrelate with 'material capabilities' and 'institutions'.

THE COMMUNICATION PROCESS AND ITS IMPLICATIONS

Virtually instantaneous transfers of billions of dollars are a daily occurrence in today's international political economy. Transnational capitalist activities are increasingly flexible as to location. There is a mounting dependency of domestic enterprises on information goods and services purchased from a limited number of TNCs. As a consequence, there is an ongoing redefinition of the physical and intellectual capacities of various political-economic agents. In sum, the emerging communication capacities of transnational corporations are significantly undercutting the 'realistic' policy alternatives of nation-states, the labour movement *and* capitalist enterprise itself. In fact, transnational communication developments may provide transnational corporations with the means to undertake a direct modification of international cultural perspectives.[34] Strategically, therefore, we need to ask questions regarding the ongoing development of *intellectual capacities* requiring a comprehensive multi-disciplinary approach.

It is on this question that the narrow conception of the mass media as 'gatekeepers' is most inadequate, in that it deflects our appreciation of intellectual capacities as products of ongoing day-to-day communication processes. In adopting this kind of approach, 'public opinion' – and, more fundamentally, intellectual capacities – can be best understood through an examination of the socialisation process, our relations with influential institutions and, most certainly, our daily activities involving material and intellectual production and consumption patterns and the characteristics of our working lives. While commercial television, for example, never functions as a purely one-way medium because its content *necessarily* reflects a consumerist view of 'reality', commercial television enables us to watch ourselves but only through the lens of man/woman 'the consumer'. As a result, television audiences, as 'consumers', are directly involved in the reproduction of their own fragmented consumerist identities.

A globalised commercial television system – now being developed through DBS and fibre optics – will afford enhanced opportunities for influencing periphery cultures. By promoting a consumerist lifestyle, transnational advertising, to quote William Melody from Chapter Two in this volume, 'can severely weaken the effectiveness of formal government policies attempting to restrict TNC access to national markets and to limit the commercialisation of their societies'. Through transnational communication developments, a new kind of core–periphery model may be emerging requiring the 'core' to be represented as the concentrated capacities of transnational corporations (TNCs) and the 'peripheries' to include a cross-section of interests, including the nation-state. Thus, as I point out in Chapter Five, while the US government's and US-based TNCs' push for an international 'free flow of information' regime constitutes an essential step towards America's long-term economic recovery (due to the relative strength and growth potential of US information and entertainment products and services), its success could well result in a significant reduction in TNCs' dependence on the United States itself. In other words, 'the eventual peripheralisation of the United States may paradoxically come about through the 'success' of US "cultural imperialism"'. The historic predominance of Hollywood's films and television programmes, as both exports and commercial models, is directly related to their status as commercial media products *par excellence*. With this in mind, DBS, for example, will quite likely contribute to the elaboration of a corporate/commercial transnational culture. Other implications and contradictions, including an elaboration of how international communication develops and its likely impact on the quantitative and qualitative aspects of transnational 'elite' identities *and* the identities and conceptualisations of the world's 'masses', are also pursued (explicitly or implicitly) in Chapters Two (by Melody), Three (by Ian Parker) and Four (by Martin Hewson).

In Martin Hewson's critique of what he calls the traditional 'institutionalist' and the Gramscian 'holist' approaches to IPE (in 'Surveillance and the Global Political Economy'), he argues that surveillance 'arises from the accumulation of information and this exercises a particular kind of influence in modern globalising society'. Surveillance, he continues, 'is the social power engendered by the accumulation of information' involving both communication media and systems of expertise. This approach is significant in that, among other things, it stresses that the locus of global power is neither the *exclusive* domain of predominant nation-states, international institutions, nor transnational class formations. Hewson argues that power also 'arises from collective practices that facilitate monitoring', involving diverse activities such as market research, census taking, international trade, cartography, and diplomacy. From this perspective, today's unprecedented global (and local) communication developments are effectively accelerating and intensifying both the monitoring functions of transnational networks and the moment-by-moment monitoring of ourselves.

In Chapter Five ('Communication Technology and International Capitalism: the Case of DBS and US Foreign Policy') – the first of three case studies in the book – I examine the role of the United States government in the development of transnational communication capacities in general and DBS systems in particular. Stephen McDowell's 'International Services Liberalisation and Indian Telecommunications Policy' (Chapter Six) provides a more detailed overview of the role of US officials in the development of an international trade regime for telecommunications equipment and services. This overview provides McDowell with the necessary political-economic context in which to analyse policy developments in India. Using propositions found in Cox, McDowell argues that the 'speed and extent to which India's telecommunications policies and economic policies have adopted a liberal model in the early 1990s is unprecedented' and these changes 'cannot be understood as arising simply from negotiation processes in which India "discovered" its interests, or simply from coercion and pressure' from US and other officials. McDowell concludes that understanding these changes requires a careful study of 'state-civil society complexes within India, as well as the context of broader transnational civil society within which the Indian state is placed'.

A similarly detailed study is provided by Milton Mueller in Chapter Seven titled 'Contested Terrain: Hong Kong's International Telecommunications on the Eve of 1997'. In this analysis of what he admits is 'an extreme case of political transition', Mueller describes a complex process involving the interactions and calculations of key countries, including the People's Republic of China, the United Kingdom and the United States; local and transnational corporate interests; and economic development assumptions regarding the appropriateness of a liberalised trade regime. Mueller concludes that the 'inevitable' liberal regime in telecommunication services is not 'inevitable' at all. In fact, Mueller's chapter locates a number of significant pre-conditions for the liberalisation of telecommunications in Hong Kong and elsewhere.

The last two chapters are in some ways the most 'thoughtful' of the collection. Adam Jones' 'Wired World: Communications Technology, Governance, and the Democratic Uprising' (Chapter Eight) acknowledges that the development of popular and accessible communication technologies – especially hand-held video cameras and fax machines – have facilitated and directly influenced recent 'democratic' uprisings in Eastern Europe, Latin America, Asia, and the peripheries of the 'developed' world (i.e. some of America's inner-cities). In explaining how these technologies have empowered people in relation to repressive state institutions, Jones understands that their application has also influenced the behaviour of both oppressors and oppressed. Moreover, as a result of the predominant role played by competitive capitalist news media and Western-trained journalists, Jones concludes that no straightforward 'equation of technology with democracy is possible'.

The final chapter by James Halloran, titled 'Developments in Communication and Democracy: the Contribution of Research', provides a critical overview of the assumptions, activities and effectiveness of social scientists engaged in international communications research over the past twenty years. In this chapter, based on his 1990 keynote presentation to the International Association for Mass Communication Research in Bled, Yugoslavia, Halloran argues that critical concerns regarding developments in communications, such as the formation and/or intensification of information/knowledge dependency relations, have not been adequately addressed in a way that might lead to feasible reforms. He then attempts to explain this failure and asks if current conditions bode well for their effective engagement. Halloran concludes that 'What we really and urgently require is a globalisation of moral responsibility' not only to redress the market-dominated model of globalisation but also to stimulate a critical, inclusive and relevant research agenda.

Halloran's concerns lead us back to William Melody's 'The Information Society: Implications for Economic Institutions and Market Theory' (Chapter Two). Here, and in Ian Parker's 'Myth, Telecommunication and the Emerging Global Informational Order: the Political Economy of Transitions' (Chapter Three), the implications of the kinds of changes described in later chapters are critically assessed in terms of their impact on the new political economic environment. The so-called 'public interest', writes Melody, is becoming increasingly identified with the interests of nationally-based telecommunications corporate 'champions'. In this context, Melody wonders if 'there is a significant difference between ITT, IBM, and Northern Telecom today, and the East India and Hudson Bay Companies of another era of colonial expansion'. As the case studies of nation-state policies in Comor and McDowell illustrate, national developments involving the rapid globalisation of telecommunication services and related activities may be characterised, using Melody's phrase, as 'an exercise in gradual self-strangulation'.

Both Melody and Parker point to a number of key contradictions involving the political-economic dynamics of transnational communication developments. Both chapters represent outstanding examples of how PEC can provide IPE with relevant, vigorously developed (but under-utilised) analytical tools. For example, Melody discusses the contradictions and inherent instabilities for the international order of current communication developments by drawing parallels to historic boom and bust cycles involving the railroad and canal developments of the nineteenth century. Melody's chapter also provides essential insights as to the strategic position of Third World countries (such as India) in relation to the rarely articulated vulnerabilities of TNCs in the current era of telecommunications-building.

Ian Parker follows with a contribution on the strategic importance of mythologies involving international telecommunication developments. Parker writes that 'changing social-economic conditions and the changes in structure of power that accompany them alter social definitions of reality and unreality, sanity and madness. The boundaries of reality are inextricably linked to the capacity to exclude the irrational, as defined principally by those with the power to enforce their definition.' Fundamental in the dissemination or distribution of such myths, therefore, is the capacity to control and put to use wealth and power. In this context, the truthfulness of myths is secondary and current global communication developments, while supported by myths such as the 'naturalness' of capitalist market activities and the 'righteousness' of the free flow of information, themselves represent the structural bases for controlling tomorrow's popular mythologies. From this perspective, the role of Cox's prospective social-economic peripheries – including the nation-state – is not only seen as being strategically important but also is provided with an approach from which communication processes may be put to effective use. Parker's chapter also provides Halloran's quest for 'a globalisation of moral responsibility' with a powerful theoretical starting point.

TOWARD A GLOBAL POLITICAL ECONOMY OF COMMUNICATION APPROACH

'Critical' writers in IPE, like Robert Cox, Stephen Gill, and others,[35] while addressing the creation of an international civil society, have not developed the heuristic and analytical tools needed to flesh out its implications and contradictions. As I have attempted to explain, while Robert Cox's work on internationalisation processes at least implicitly recognises the saliency of PEC concerns, it very much requires a more developed theory of the communication process. As Rob Walker has written at a more general level, 'fairly profound transformations are currently in progress. But ... our understanding of these transformations, and of the contours of alternative political practices, remains caught within discursive horizons that express the spatiotemporal configurations of another era.'[36]

With technological, political, economic and cultural roads now being cleared towards a world characterised by instantaneous transnational communications, one of the central dynamics shaping the political economic activities of advanced industrialised countries and transnational corporations is the flowering of unprecedented capacities to control the technological, political, economic and cultural lives of all human beings. In addressing these developments, a PEC perspective – as represented in this collection and elsewhere – constitutes an invaluable base on which to develop required heuristic and analytical tools. IPE scholars, while recognising the importance of communication, have rarely addressed the subject directly. Taken as a whole, the following chapters represent just one building block in the construction of a much-needed bridge among students of the two fields. This book and this bridge – as communication media – seek to stimulate further debates within and between these complementary perspectives. As Harold Innis recognised, 'the subject of communication offers possibilities in that it occupies a crucial position in the organization and administration of government and in turn of empires and of Western civilisation'.[37]

Notes

1. Thanks to Martin Hewson and J. Magnus Ryner for their helpful comments on earlier drafts of this introduction.
2. Robert W. Cox, 'Social Forces, States and World Orders: Beyond International Relations Theory', in *Millennium: Journal of International Studies*, Vol.10, No.2 (Summer 1981) p.126.
3. Ibid.
4. Statistics from Howard H. Frederick, *Global Communication and International Relations* (Belmont, CA.: Wadsworth, 1993) pp.58–9.
5. The only chapter in this volume that has been published elsewhere is William Melody's contribution in Chapter Two.
6. See, for example, the writings of Stuart Hall, David Morley and Raymond Williams, among others.
7. This phrase was popularised through the widely read book by Peter L. Berger and Thomas Luckmann, *The Social Construction of Reality, A Treatise in the Sociology of Knowledge* (Garden City, NY: Anchor Books, 1967).
8. For example, see David Morley, 'Where the Global Meets the Local: Notes from the Sitting Room' in *Screen*, Vol.32, No.1 (Spring 1991) pp.1–15.
9. See, for example, writings by Robert Babe, Nicholas Garnham, William Melody, Ian Parker, Vincent Mosco, Herbert Schiller, and many others. An excellent exegesis on the thought of Gramsci and its applicability to the general concerns outlined here is in Enrico Augelli and Craig Murphy, *America's Quest for Supremacy and the Third World, A Gramscian Analysis* (London: Pinter Publishers, 1988).
10. The overviews of the realist and the liberal complex interdependence perspectives discussed below are based on those provided in Stephen Gill, *American Hegemony and the Trilateral Commission* (Cambridge: Cambridge University Press, 1990) pp.11–25.
11. Cox, quoted in Stephen Gill, 'Epistemology, Ontology and the "Italian School"' in Stephen Gill (ed.), *Gramsci, Historical Materialism and International Relations* (Cambridge: Cambridge University Press, 1993) p.42.
12. Ibid., p.43.
13. See Robert G. Gilpin, 'The Richness of the Tradition of Political Realism' in *International Organisation*, Vol.38, No.2 (Spring 1984) pp.287–304.
14. See Robert O. Keohane and Joseph S. Nye, *Power and Interdependence* (Boston: Little Brown, 1977).

15. See John G. Ruggie, 'International Regimes, Transactions and Change – Embedded Liberalism in the Post-War Economic Order' in *International Organisation*, Vol.36 (1982) pp.379–415.
16. Robert Cox, 'Gramsci, Hegemony and International Relations', p.171.
17. Gramsci quoted in Robert Cox, 'Social Forces, States and World Orders' (revised version) in R.O. Keohane (ed.), *Neorealism and its Critics* (New York: Columbia University Press, 1986) p.161.
18. Of course, as heuristic tools, these 'categories of forces' are interchangeable. The international banking system, for example, can be categorised as either/both an Institution or/and a Material Capability.
19. Structures, as ideal types, are applied to three levels of human activity: (1) the organisation of production involving the application of 'Social Forces' (as in the structuring, for example, of a Fordist or post-Fordist accumulation process); (2) state/society complexes and their formal expression as 'Forms of State'; and (3) the ensemble or configuration of forces and state/society complexes in the form of 'World Orders'. In sum, the three broad categories of forces as represented in Cox's dialectical triad are themselves dialectically related to the development of a particular configuration of world structures.
20. Robert W. Cox, *Production, Power, and World Order. Social Forces in the Making of History* (New York: Columbia University Press, 1987) p.7.
21. Ibid., p.321–7.
22. Ibid., p.224.
23. Ibid., pp.244–51.
24. Ibid., p.253.
25. Ibid., pp.253–4.
26. Of course Cox does discuss factors impeding the organisational capacities of workers. See for instance, 'Social Forces, States and World Orders: Beyond International Relations Theory' in *Millennium: Journal of International Studies*, Vol.10 (1981) pp.148–9.
27. Ibid., p.130.
28. Strange, quoted in Cox, *Production, Power, and World Order*, p.257.
29. Strange, quoted in Ibid.
30. In *Production, Power, and World Order*, Cox places great weight on the role of institutions in the ongoing formation of ideological consensus, but fails adequately to elaborate on the process itself. See for example p.259.
31. Ibid., p.294.
32. In 'Gramsci, Hegemony and International Relations: An Essay in Method' in *Millennium: Journal of International Relations*, Vol.12, No.2 (Summer 1983) (at p.173), Cox points to how the notion of 'self-reliance' was effectively transformed into *meaning* something complementary to the world hegemonic order. For example, when asked about US economic aid to less developed countries, Ronald Reagan's frequently repeated fisherman analogy ('if you give a man a fish today he'll go hungry tomorrow; but if you teach a man how to fish [ie. how to function under a 'free' market regime], he'll never go hungry again ...') was employed in the process of transforming Third World demands for endogenously-determined development into something altogether different.
33. Robert Cox, *Production, Power, and World Order*, p.299.

34. On this general argument, see Herbert I. Schiller, *Culture, Inc., The Corporate Takeover of Public Expression* (New York: Oxford University Press, 1989).
35. These 'others' include Enrico Augelli, Craig Murphy, Bruce Russett, Susan Strange and Kees Van Der Pijl.
36. R.B.J. Walker, *Inside/Outside: International Relations as Political Theory* (Cambridge: Cambridge University Press, 1993) pp.ix–x.
37. Harold A. Innis, *Empire and Communications* (Toronto: University of Toronto Press, 1972) p.5.

Part I
Critical Perspectives

2 The Information Society: Implications for Economic Institutions and Market Theory[1]

William H. Melody

An ever-more-popular theme in the social sciences, as well as in the general literature over the past decade, has been that technologically advanced economies are in the process of moving beyond industrial capitalism to information-based economies that will bring profound changes in the form and structure of the economic system.[2] Rapid advances in computer and telecommunication technologies are making it possible to generate information that was heretofore unattainable, transmit it instantaneously around the globe, and – in a rapidly growing number of instances – sell it in information markets. Some authors claim that the United States already devotes the majority of its economic resources to information-related activities.[3] The computer, telecommunication, and information content industries are among the most rapidly growing global industries, and are expected to remain so for the next decade. Many national governments are counting on these industries to provide the primary stimulus to their future growth.

Economists are now just beginning to recognise that the most important resource determining the economic efficiency of any economy, industry, productive process, or household is information and its effective communication. The characteristics of information define the state of knowledge that underlies all economic processes and decision-making structures. Fundamental changes in the characteristics of information, and in its role in the economy, should be central to the study of economics, but as yet they are not. The state of information in the economy has pervasive effects on the workings of the economy generally. It has intensified impacts on those sectors that provide information products or services, for example, press, television, radio, film, mail, libraries, banks, credit bureaus, data banks, and other 'information providers', as they are now called.[4] And the establishment of information markets brings about changing conceptions of public and private information, as well as the property rights associated with marketable information.[5] In this chapter I examine some of the significant implications of the changing role of information for economic institutions generally and for market theory.

TECHNICAL EFFICIENCY AND MARKET EXTENSION

An expansion of available information, together with enhanced and improved telecommunication, should permit more efficient decision-making and the extension of markets across geographical and industry boundaries. It should increase competition. It should allow resources to be allocated more rapidly and efficiently. The conditions of real markets should approximate more closely the assumptions of theory, where markets are frictionless and operate under conditions of perfect information. Indeed much of the literature on the information economy considers these developments to provide unmitigated benefits to society.[6]

But closer examination indicates that the benefits of these technologies will not be distributed uniformly across markets, that certain segments of society will be made poorer both in absolute as well as relative terms, and that the structure of markets in many industries will be affected in fundamental ways.[7] These new technologies permit markets to be extended to the international and global level. But only the largest national and transnational corporations (TNCs) and government agencies have the need for, and the ability to, take full advantage of these new opportunities. For them the geographic boundaries of markets are extended globally, and their ability to administer and control global markets efficiently and effectively from a central point is enhanced.

The manner in which these technological developments are being implemented creates a significant barrier to entry for all but the largest firms, thereby accelerating tendencies toward concentration.[8] In fact smaller firms are likely to find themselves disadvantaged because of the new technological developments. For example, the telecommunication systems in Canada and other technologically advanced countries are being redesigned to meet the technically sophisticated digital data requirements of high volume, multiple purpose, global users. For traditional, simpler communication requirements, such as a basic telephone service, the new upgraded system will serve quite well, but at substantially increased costs to smaller users.[9] The telecommunication options available to small, localised, and even regionalised businesses do not reflect their unique needs. Rather, their range of choice is dictated by the national and global needs of the largest firms and government agencies. The most efficient telecommunication system for their needs has been cannibalised in the creation of the technologically advanced system.

In most industries the new competition is simply intensified oligopolistic rivalry among TNCs on a world-wide basis. The firms that can now leap across market boundaries are already dominant firms in their respective product and geographic markets. Their entry has a major impact on the structure of the supply side of the market and prompts a strategic response from the established dominant firm(s). This is not atomistic competition responding to market forces that reflect consumer demand, but rather a type of medieval jousting for territorial control.

The focal point of this oligopolistic rivalry is on differentiated adaptations of particular technologies and product lines for sale to nation-states. Major decisions involving multi-million-dollar commitments over many years are made relatively infrequently, for example, selecting a satellite system or a line of computer or telecommunication equipment. The rivalry is directed to obtaining a long-run position of market entrenchment and dominance in particular foreign national submarkets.

The rivalry among TNCs for entrenchment in new national markets differs fundamentally from traditional market theory in several respects. First, short-run market clearing prices are not the focal point of the rivalry. Rather, short-run pricing policy is simply one of many strategic tools for achieving the long-run objective. This rivalry stands far outside the short-run pricing behaviour examined by traditional oligopoly theory.

Second, competitive advantage is obtained not primarily from the superiority of a product in the eyes of individual consumers exercising choice, but rather from effective persuasion of government leaders in foreign countries. The objective is to secure a position of special privilege in entering national markets. The privileged market position then is ensured by the national policy of the purchasing countries with respect to such matters as licensing, tax, tariff, currency exchange, capital repatriation, entry barriers imposed on rivals, etc.

In attempting to achieve these long-term dominant market positions, the TNCs are assisted by governments of their respective home-base countries. The home governments adopt policies and positions that will assist their respective TNCs, and sometimes they even participate in institutional marketing. Thus, the oligopolistic rivalry among TNCs involves a strong element of nationalism and direct government involvement on both the demand and supply sides of the market exchange.

Adoption of the new technologies tends to increase the significance of overhead costs, not only for the information and telecommunication activities, but also with respect to greater centralisation of functions and capital/labour substitution, for example, robots. Thus, the inherent instability in oligopoly markets is magnified by the instability created by an increased and very significant proportion of overhead costs.

Taken collectively, these changes introduce new elements of risk. But they also provide new opportunities to shift these risks away from TNC investors and managers to the particular localities where production occurs, and the institutions that reside there, that is, local government, labour, and consumers. TNCs also can diversify their risks by expanding their absolute size and geographical coverage. The larger the TNC, the more resources at its command for allocation within the firm rather than through capital markets. The greater the geographical coverage, the more risks can be diversified by the TNC, although these risks could be disastrous for any particular production location dependent on the TNC. In addition, the enhanced market power strengthens the TNC's ability to exploit both resource and consumer markets.

Because the new technologies permit rapid transfer of new types of information, they permit more frequent short-run decision-making by TNC managers. In global markets, the terms of trade, currency exchange rates, interest rates, and money movements are often as important to real profitability as the actual production of goods and services. With new opportunities for frequent, short-run decisions there is likely to be an increased emphasis on day-to-day financial transfers, if not ongoing speculative manipulation. This will create additional instability for any particular resource supplier or production location that might fall out of favour as a result of short-term shifts in financial and currency markets.

Historically, the current revolution in telecommunication technology can be compared in certain respects with the effect of the introduction of the telegraph upon the structure of markets in the United States over the period 1845 to 1890. In his study of these developments, Richard DuBoff concluded: 'The telegraph improved the functioning of markets and enhanced competition, but it simultaneously strengthened forces making for monopolisation. Larger scale business operations, secrecy and control, and spatial concentration were all increased as a result of telegraphic communications'.[10] In fact, he says, 'increasing market size helped "empire builders" widen initial advantages which at first may have been modest'.[11] DuBoff's assessment provides a useful benchmark for examining the current global developments that illustrate similar economic effects, but substantially magnified and modified as described in this chapter.

THE ROLE OF NATIONAL GOVERNMENTS

For the TNCs, the domestic markets in their home countries provide a springboard to their activity in global markets. The home governments identify more directly with the international success of particular TNCs because they play an important role in the home country's domestic economy.

Home governments tend to exhibit greater tolerance for increased domestic monopoly power because it enhances the power of their resident TNCs in international markets. This can range from a reduced emphasis on the application of anti-trust and anti-combines laws to the actual encouragement of domestic cartels. In the United States, it will be recalled, the Webb–Pomerene Act has provided anti-trust immunity for foreign markets for seventy years. Some countries, for example Canada and some Western European nations, have promoted domestic monopoly power in some industries for the purpose of creating a larger corporate presence that they hope will have the power to compete with the largest TNCs, mostly based in the United States.

Today, the US government is extremely concerned about the access of US-based TNCs to foreign markets. Government advocacy of free trade is designed to open markets for some industries, particularly the computer, telecommunications, and information-content industries. Restrictions on Japanese automobile imports illustrate a comparable defensive reaction.

As oligopolistic rivalry in global markets becomes more intense, national governments are more actively attempting to manipulate the terms of international rivalry to the advantage of 'their' TNCs. Thus, the TNCs are becoming more direct instruments of macro-economic policy through research and development (R&D) subsidies, tax concessions, tariff conditions, trade agreements, and other policies. This includes the assumption of market risk by home governments in the form of R&D funding, investment guarantees, government–industry joint ventures, and government assistance of home-based TNCs in international market negotiations through applying political pressure to foreign governments.

Today this kind of government involvement is labelled 'industrial policy'. For most nations, some kind of industrial policy has been a prerequisite for survival for a long time because their economies have been so dominated by foreign TNCs and larger, more powerful nations. The current interest in industrial policy in the United States is in part a reaction to this, but is also influenced by a decline in US dominance in some global markets as well as a reaction to the prolonged recession in the US domestic economy.

This new approach reflects a change in the role of government from adopting policies designed to stimulate the marketplace environment generally, toward adopting more focused policies designed to assist specific companies. As such, it reflects direct interference in the market. It identifies the economic prosperity of the nation with the financial success of the largest home-based TNCs. The role of government then becomes one of using its political power to manipulate the rules by which the market works to the advantage of the TNCs that it has chosen to support. In Canada, this is called 'picking winners'. These policies provide significant barriers to entry to those firms not selected, which includes all domestic firms not large enough to exploit international markets.

Under these conditions competition in the domestic market can be seen as potentially damaging to the ability of home-based TNCs to compete successfully in global markets. Anti-trust policies and pro-competitive domestic policies become less important, if not antiquated. Monopoly and cartel behaviour are accepted as tolerable, if not promoted. Even monopolistic exploitation of domestic consumers becomes tolerable as providing the necessary strength, power, and resources to compete successfully in the global markets.[12] Politically it is much easier to provide subsidies by simply allowing a home-based TNC to exploit monopoly power in the domestic market rather than going through the cumbersome political process of first taxing and then granting subsidies.

In this new political economic environment, the conception of the public interest within a nation also changes. Traditional concerns about the prices and quality of public utility services and the universality of coverage of public service declines. For example, in the United States, basic telephone service as a priority of social policy is being questioned, if not yet abandoned, by the Federal Communications Commission (FCC). The international success of home-based TNCs, as measured by sales, profits, and a favourable balance of payments, becomes a primary objective of government public policy. This success is viewed as fueling domestic employment, productivity, and national wealth. Domestic consumers and social policies are seen to benefit from the trickling-down of benefits from successful TNCs. How social services will be funded from this wealth accumulated by TNCs, when government policy is directed to subsidising their competitive efforts in global markets rather than taxing away their monopoly profit, remains a mystery.

The real change is a much closer identification of the national and public interests with the corporate interest of the dominant home-based TNCs. It is truly ironic that these industrial policies typically are justified by invoking the ghost of Adam Smith and free market competition. In fact, they represent a fundamental mistrust of the free market. We need not be reminded that, in fact, Smith argued that the wealth of nations would be enhanced if domestic competition were encouraged rather than sacrificed to the myopic criterion of mercantilistic success.[13] One might question whether, under this approach, there is a significant difference between ATT, IBM, and Northern Telecom today, and the East India and Hudson Bay Companies of another era of colonial expansion.

In addition, national government policy designed to promote the power of TNCs in global markets may well be an exercise in gradual self-strangulation. The nature and direction of government policy intentions are always heavily constrained by market conditions and the power of corporations and other large economic units to prevent their effective implementation. Canada has been attempting to implement independent economic and cultural policies for generations. But it has neither the economic nor the political power to implement them effectively in the face of domination by US-based TNCs. As national governments tie themselves more closely to promoting the corporate power of their TNCs, they are at the same time reducing their own degrees of freedom to adopt domestic or international policies contrary to TNC interests.

THE ROLE OF MARKET THEORY

Traditional market theory provides a perfect rationale for this expansion of TNC market power. By assuming that technology is autonomous and beneficial, that oligopolistic rivalry for long-term dominant positions in foreign national markets is competition, and that market-clearing prices maximising short-run profit will yield optimal resource allocation, the theory simply reflects the short-run market power positions of the dominant firms. Such concepts as static equilibrium, marginal cost, and consumer surplus are perfectly pliable in their subjective application. Indeed, within this theoretical framework, nothing can be rejected that travels under the appropriate theoretical labels.

More recent theoretical developments such as Ramsey pricing, sustainability, and contestability move a step further away from reality into abstract metaphysics.[14] Under these new theoretical developments, one judges the effectiveness of competition in markets not by the existence of real competitors, but by the possibility that there might be. In keeping with most of neoclassical theory, alternatives not followed provide the sole basis for judging the efficiency of actual market conditions. It requires only a modest imagination to hypothesise an alternative set of conditions that would be worse (or better) than the reality experienced.[15] Either way, it is not very helpful in solving the problems of efficient resource allocation in the real world.

A current illustration of the abuse of market theory in the US telecommunication industry is now unfolding in policy debates before the FCC, Congress, and more recently state regulatory commissions and legislative committees. The regulated telephone companies suddenly have discovered that the costs associated with the local loop facilities connecting telephones and other terminal devices to the central office, as well as certain central office functions, are not sensitive to variations in the volume of traffic – that is, they are short-run fixed costs. In addition, these costs are common to multiple services, including local, long distance, and specialised data services. It is argued that neoclassical theory does not permit the allocation of any short-run common costs to competitive (or contestable) long distance or data services. If high charges for accessing the system are imposed upon large business users, it is claimed that some will bypass the telecommunication network.

But these same fixed common costs apparently can be allocated to the monopoly basic telephone service. High access charges for local telephone service may force a good many poor people to give up telephone service. But that is tolerable because the estimated consumer surplus gains from increased long distance use by large-volume users will exceed the consumer surplus losses of disconnected customers with no alternatives they can afford.

The point here is that neoclassical market theory can justify equally well virtually any result. What it does tend to reflect in reality is the existing distribution of market power. The common interest among local telephone companies, long distance carriers, and large industrial users has permitted them to purchase a version of the theory that reflects that interest.

This version of the theory ignores the fact that the cause of increased costs of network access is upgrading the local exchange to meet the requirements of digital data and transcontinental communication, essential to enhanced data services and global networks, but unnecessary for local telephone service. It also ignores the fact that benefits to society do not decline when a business shifts some of its telecommunication traffic from a telephone company and begins using an alternate source of supply. However, societal benefits do decline, not only to that subscriber but also to others who would call that subscriber, when a subscriber is forced to disconnect service. Moreover, disconnected subscribers save the telephone company little, if any, costs. In the vast majority of instances, disconnections will be in the poorer parts of town where the telephone company has no alternative use for its loop investment. It will be stranded. In my view, a correct interpretation of neoclassical theory would require that no subscribers be permitted to disconnect unless they were unwilling to pay a price equal to their respective short-run social marginal costs. For users already connected, this must be close to zero, if not negative. However, local residential customers are not in a position to mount a competitive lobby advocating neoclassical theory interpreted from the perspective of their interest. Selection of the 'appropriate' interpretation of neoclassical theory is not determined by an independent analysis of facts, but by the power of the interest groups in advocating rival interpretations.

Finally, it should be noted that one influential argument for reducing rates for long distance voice and data services at the expense of local telephone rates is to help domestic companies (in the United States and Canada, at least) compete more effectively in national and global markets. In the not-too-distant future local telephone rates may include a hidden cross-subsidy to give home-based TNCs an artificial competitive advantage in seeking foreign contracts.

The most relevant market model for examining the consequences of competition in the information age is one of indeterminate, unstable oligopoly wherein the TNCs deliberately employ short-run pricing strategy to achieve long-run entrenchment and monopoly power in national markets, foreign and domestic. For detailed analytical development of this type of market model, one must look at Joseph Schumpeter, Karl Marx, and other institutionalists following in the same tradition. Within this oligopoly model, the market provides ample room for negotiation to affect outcomes in both the short run and the long run, with a wide range of possibilities. Therefore attention must be paid to negotiating structures, criteria and alternatives, an area of analysis that has not been well developed.

The new oligopoly markets cannot be explained without reference to dependency theory. Incorporation of the possibility of dependent market relations simply recognises that buyers and sellers are not part of a unified homogeneous market. The locational separation of economic functions may be total. The source of resources, production, consumption, profit recognition, and control over continuing reallocations of resources may each be in a different country. Many localities are dependent on a specialised production plant of a TNC used to serve markets on another continent. In fact, different economic, political, social, and cultural systems generally provide a basis either for significant specialised advantage to a TNC, or a significant barrier to production and marketing. In the 'information age,' dependency relations within global markets take on a new significance.

IMPLICATIONS

Dependency theories have some significant differences.[16] But they also are characterised by some common conditions.[17] Control is exercised at the market centre. Peripheral or hinterland locations are developed primarily to serve the interests of the major centres of power by exploiting natural resources, low labour costs, or other elements of specialised, comparative advantage. Effective control over the type, the direction, and the rate of development rests at the centre. Thus, the periphery is dependent on the centres of power. Apparent short-run efficiency and stability in the central markets is obtained at the expense of instability and distorted development in the outlying areas. The economy in the region bears the risk both of changes in short-run market conditions at the centre and of the potential loss of its specialised comparative advantage to another region. But even this market arrangement cannot protect the centre market from long-run instability, and under some circumstances may even accentuate it by requiring less frequent but more severe adjustments. Harold Innis concluded from his studies that 'the economic history of Canada has been dominated by the discrepancy between the centre and the margin of western civilisation'.[18]

As TNCs expand using the information technologies, they can reduce their dependency upon any single resource supply or production location, thereby enhancing their negotiating power with individual governments, unions, and other groups. A higher proportion of risk can be transferred to the resource supplier and producing regions. This can be done by means of pressure:

1. For subsidies, tax concessions, and regulations conferring special privileges or even government promotion of TNC interests.
2. For the maintenance of a labour force of specialised skills at low wages in the face of unstable employment.
3. For exemption from social controls such as health and environmental standards.
4. For a privileged position in the domestic market of the peripheral producing nations.

In addition, the TNCs may well be able to pass on an additional share of risk to their respective home countries by negotiating for special privileges. Thus the information technologies become a major tool for TNCs to enhance both their market power and their autonomy in negotiating with nation states, whether hinterland producers or home governments. Moreover, with direct broadcast satellites, and the massive spillover of television signals across national borders, TNC mass media advertising can severely weaken the effectiveness of formal government policies attempting to restrict TNC access to national markets and to limit the commercialisation of their societies. New models of dependency may have to consider TNCs at the centre, with all affected nations relating to them from different positions of dependency.

A significant separation of the incidence of risk from the beneficiaries of risk can have severe implications. It is an invitation for the TNCs to assume risks far beyond the level they would assume under normal market conditions or rational resource allocation. This may explain the willingness of American, Canadian, and European banks in 1981 to violate their own lending rules in loans to Third World nations in an attempt to exploit the very high short-term interest rates. Given the fact that oligopolistic rivalry for long-run entrenchment in national markets is a game with very large pay-offs, the incentive to over-invest both in production facilities and in market creation activities is great. When placed in a context of high overhead costs and fluctuating demand, the inherent instability in the economic system is likely to be magnified significantly.

Under these conditions, there is no reason to expect that the price system will work toward long-term efficiency in resource allocation. Short-term prices are likely to fluctuate significantly over wide ranges reflecting either excess capacity, short-term shortages, or opportunities for monopolistic exploitation. The market provides no constraint on over-investment. Indeed it would appear that the boom and bust cycles of investment in railroads and canals more than a century ago may provide the best historical reference point for analysis of the forthcoming surge in the computer, telecommunications, and information content industries.

After his study of the history of the megaproject investments of the past, Innis concluded: 'As a result of the importance of overhead costs, in its effects on inelastic supply and especially joint supply, the price level has become an uncertain and far from delicate indicator in adjusting supply and demand'.[19] Recent experiences with nuclear power, oil exploration, and international banking do not provide a basis for confidence that the economic system functions any better now in this respect than it did a century or two ago. Certainly a relevant model for the analysis of future global markets is the oil market, but with TNCs in the role of OPEC.

Clearly, if instability in the global economic system increases, it will affect all parties. Thus, the TNCs, by having the power to transfer the incidence of risk, may be stimulated to engage in investment and pricing behaviour that will create more instability and risk than they are able to transfer. This is consistent with the standard explanations of instability from neoclassical oligopoly theory.

In this market environment neoclassical price theory is somewhat akin to a set of decision rules for optimizing the arrangement of the deck chairs on the *Titanic*.[20] The price signals tell us nothing about the speed and direction in which the economic ship is headed. Short-run prices are likely to be very misleading as guides to resource allocation. Here one cannot help recalling a passage from C.F. Ayres:

> The time will come when we shall see that the root of all our economic confusion and the cause of the intellectual impotency which has brought economics into general disrepute is the obsession of our science with price theory – the virtual identification of economics with price analysis in almost total exclusion of what Veblen called the 'life process' of mankind.[21]

The challenge for economic analysis is to address directly the problem of long-term resource allocation under conditions of unstable oligopoly, and the implications of policies designed to contain its worst possibilities.

These developments obviously do not bode well for the maintenance of traditional notions of public or consumer interests in domestic markets. Existing government regulatory agencies are likely to become even less effective in the future than they are now. Yet, the need for advocacy of consumer interests will be greater. Public interest advocates will have to seek other avenues of representation through the legislative and judicial process.

Countries such as Canada find themselves ambivalent with regard to these developments in the information industries. Canada has spent most of its history striving to maintain a small degree of independence in the shadow of first Britain and then the United States. But Canada is at the forefront of telecommunication technology and sees a window of opportunity to exploit global information and telecommunication markets.

Canada's major entry into the global information market sweepstakes – as policy makers sometimes call it – is Bell Canada Enterprises (BCE), most frequently through its equipment manufacturing subsidiary, Northern Telecom. In the early 1980s Bell Canada undertook a corporate reorganisation in which it created the holding company, BCE, and moved all of its activities except domestic telephone service out from under government regulation. This was necessary, according to the company, to position itself for competing in global markets against larger TNCs like ITT, ATT, Siemens, Phillips, IBM, and so on. BCE has won lucrative contracts in

Saudi Arabia and other developing countries, and is currently concentrating on China. The government has yet to get around to passing legislation that will give its regulatory agency, the Canadian Radio-Television and Telecommunications Commission (CRTC), effective power to investigate possible cross-subsidies between Bell Canada's regulated and unregulated activities. Recently Bell Canada announced that it will be seeking to rebalance its rates, following the access charge pattern already begun in the United States. It appears that success in the global market has taken priority over domestic social policy.

BCE has been sufficiently successful in the global telecommunication market that it has opened plants throughout the world. BCE employment in Canada has declined in recent years and rumours persist that the company may shift the headquarters of its successful equipment subsidiary to the United States. The Canadian government undoubtedly is seeking ways to keep BCE as a successful Canadian TNC. But BCE's interest seems to be shifting further and further from that of the Canadian domestic economy for any purpose other than exploitation.

It would appear that Third World nations will bear the brunt of the risk and instability associated with the exploitation of information industry technologies and markets. As producers in the periphery, they will have little if any control over the product or profit from their labour and other resources. Moreover, successful global marketing by the TNCs requires that Third World leaders be convinced of the need to import the latest computer/telecommunication systems. The Third World represents an extremely significant part of the potential market. But if they buy into these technologies, most will be committed to a long-run dependence that will contribute to continuing short-term balance of payments problems, and virtually permanent constraints on their domestic economies.

For nations with a 50 per cent literacy rate, a 5 per cent telephone penetration rate, and an average annual per capita income less than the cost of a computer terminal, one might question the priority of new technologies that provide instant access to data banks in New York and London, except of course that they facilitate TNC control in the region. In the future, Third World nations desperately need to establish effective negotiating strategies to resist the technological salesmanship of the TNCs and their home governments, and to attempt to assert their own priorities. In a global market of oligopolistic rivalry, there is negotiating room for smaller countries to establish a variety of new co-operative and competitive market relations. In some areas the non-aligned nations have already begun to take modest steps in this direction. The relative negotiating positions may be unequal, because of differences in economic and political power, but they are not dictated by market structure. The inherent instability of rivalrous oligopoly behavior provides a soft underbelly to the power of dominant TNCs. If, for example, the Third World decided not to buy into the new information technologies, there would be at least instability, and possibly a short-run collapse in these industries at the TNC centres.

CONCLUSION

Although this analysis of some of the implications of the information economy is not comforting, neither is it entirely bleak. The developments outlined in this chapter will bring to the foreground and accentuate the oligopolistic character of most national and global markets. In making the reality more visible, the problems may be addressed more directly. Perhaps it will provide a stimulus to reorient market theory from abstract notions of atomistic competition to the challenging reality of indeterminate, unstable oligopolistic rivalry.

Whereas the competitive model invites the abdication of analysis of real markets, the oligopoly model demands it. Significant progress will be made only when economics examines more deeply such matters as:

1. The characteristics of equal and unequal exchange;
2. The dimensions of oligopostic negotiation;
3. Criteria for determining long-run balanced growth;
4. The specific causes of market instability;
5. The real-world opportunity-cost characteristics of overhead functions;
6. Long-run efficiency criteria for sharing common costs in multi-product firms;
7. The rules and regulations essential to the efficient functioning of different kinds of markets;
8. Mechanisms for accountability with respect to actual resource allocation in real markets;
9. Comparisons between market and non-market systems of resource allocation;
10. Methods for incorporating externality and public interest considerations in long-run resource allocation decisions.

There are unlimited opportunities for institutional analysis and theoretical developments to make complementary contributions. Perhaps the changes unfolding in the information economy will provide the catalyst.

Notes

1. This chapter is a modified version of a paper under the same title originally published in June 1985. Reprinted from the *Journal of Economic Issues* (Vol.XIX, No.2) by special permission of the copyright holder, the Association for Evolutionary Economics.
2. See for example Daniel Bell, *The Coming of Post-Industrial Society: A Venture in Social Forecasting* (New York: Basic Books, 1973) and Marc Porat, 'Global Implications of the Information Society,' *Journal of Communication*, Vol.28 (Winter, 1978) pp.70–80.
3. Fritz Machlup, *Knowledge: Its Creation, Distribution, and Economic Significance*, 3 vols (Princeton, NJ: Princeton University Press, 1980–84); Porat, 'Global Implications'.
4. See for example, Dallas W. Smythe, *Dependency Road: Communications, Capitalism, Consciousness, and Canada* (Norwood, NJ: Ablex, 1981); and William H. Melody, 'Direct Broadcast Satellites: The Canadian Experience' (1982), published in German in *Satelliten-Kommunikation: Nationale Mediensysteme und Internationale Komrnunikationspolitik* (Hamburg: Hans Bredow Institute, 1983).
5. Herbert I. Schiller, *Who Knows: Information in the Age of the Fortune 500* (Norwood, NJ: Ablex, 1981); and Rohan Samarajiwa, 'Information and Property Rights: The Case of the News Agency Industry' (Paper presented at 14th Conference of the International Association for Mass Communication Research, Prague, Czechoslovakia, August 1984).
6. See for example Ithiel de Sola Pool, *Technologies of Freedom* (Cambridge, Mass.: Harvard University Press, 1983).
7. William H. Melody, 'Development of the Communication and Information Industries: Impact on Social Structures' (Paper prepared for the *Symposium on the Cultural, Social, and Economic Impact of Communication Technology*, sponsored by UNESCO and Instituto della Enciclopedia Italians, Rome, Italy, 12–16 December 1983).
8. Edward S. Herman, *Corporate Control, Corporate Power: A Twentieth Century Fund Study* (New York: Cambridge University Press, 1981).
9. William H. Melody, 'Cost Standards for Judging Local Exchange Rates,' in *Diversification, Deregulation, and Increased Uncertainty in the Public Utility Industries*, ed. H.M. Trebing (East Lansing, Mich.: Michigan State University, MSU Public Utilities Papers, 1983) pp. 474–95; 'Efficient Rate Regulation in the Competitive Era', in *New Directions: State Regulation of Telecommunications* (Symposium Proceedings, Washington State Legislature, Joint Select Committee on Telecommunications, and University of Washington Graduate School of Public Affairs, Seattle, 11–12 July 1984, Sect.VI, pp.1–18).
10. Richard B. DuBoff, 'The Telegraph and the Structure of Markets in the United States, 1845–1890' in *Research in Economic History*, Vol.8 (1983) pp.253–77.
11. Ibid., p.270, footnote omitted.
12. See, for example, Robin E. Mansell, 'Industrial Strategies and the Communication/Information Sector: An Analysis of Contradictions in

Canadian Policy and Performance' (Ph.D. diss., Simon Fraser University, 1984); and 'Contradictions in National Communication/Information Policies: The Canadian Experience', *Media Culture and Society* (Spring 1985) pp.33–53.

13. See Adam Smith, *An Inquiry into the Nature and Causes of the Wealth of Nations*, 5th ed. (New York: Modern Library, 1977 [1776]).
14. See, for example, William J. Baumol, 'Contestable Markets: An Uprising in the Theory of Industry Structure' in *American Economic Review*, Vol.72 (March 1982) pp.1–15; for a critique, see William G. Shepherd, '"Contestability" vs. Competition' in *American Economic Review*, Vol.74 (September 1984) pp.572–85.
15. See William H. Melody, 'The Marginal Utility of Marginal Analysis in Public Policy Formulation', in *Journal of Economic Issues*, Vol.8 (June 1974) pp.287–300.
16. Gabriel Palma, 'Dependency: A Formal Theory of Underdevelopment or a Methodology for the Analysis of Concrete Situations of Underdervelopment' in *World Development*, Vol.6 (July/August 1978) pp.881–924.
17. See, for example, Harold A. Innis, *Empire and Communications* (Toronto: University of Toronto Press, 1972 [1950]); Harold A. Innis, *Essays in Canadian Economic History* (Toronto: University of Toronto Press, 1956); and I. Wallerstein, *The Modern World System: Capitalist Agriculture and the Origins of the European World Economy in the Sixteenth Century* (New York: Academic Press, 1976).
18. Innis, *Essays in Canadian Economic History.*
19. Ibid., p.130.
20. The analogy is not mine. Source unknown.
21. C.F. Ayres, *The Theory of Economic Progress: A Study of the Fundamentals of Economic Development and Cultural Change* (Michigan: New Issues Press, Western Michigan University, 1978 [1944]), p.xxxv.

3 Myth, Telecommunication and the Emerging Global Informational Order: The Political Economy of Transitions

Ian C. Parker

> Enormous improvements in communication have made understanding more difficult.
>
> (Harold Innis[1])

> A warlike, various and tragicall age is the best to write of and the worst to write in.
>
> (Abraham Cowley 1651)

> Reality: what a concept!
>
> (Robin Williams 1979)

This chapter focuses on future developments in the telecommunications sector, with some attention to the role of telecommunication satellites in these developments. Its perspective, method, and conclusions, however, do not rest on the assumption that telesatellites are the predominant medium of telecommunication as we enter the twenty-first century, although they are one of the more powerful and significant new technologies. Rather, it is concerned to suggest that telesatellites gain their full political-economic significance as a major component in a number of sets of actual and potential systems of global communication involving many other media, in a period when there is still considerable uncertainty about which media-systems and configurations are likely to predominate in the early twenty-first century. Telesatellites are a strategic component of a period of transition which is not defined exclusively by the presence of telesatellites.

The argument of the chapter may be summarised in terms of the following propositions:

1. The period from about 1945 to the early twenty-first century may legitimately be regarded as a period of major global political-economic transition.
2. In periods of major transition, levels of uncertainty (as defined below) tend to be relatively high, and conventional mainstream social-scientific models which presuppose a relatively constant structure of social-economic relations tend to decline in relevance.
3. In such transition periods, myths (in the sense defined below) are essential strategies of uncertainty-reduction. The demand for myths increases in periods of heightened uncertainty. Telecommunications and other, frequently related, developments since 1945 have contributed to a heightening of uncertainty, and hence to an increased demand for myths, including myths of telecommunications.
4. Most present myths of telecommunications and of the emerging global informational order involve serious limitations, as well as important partial truths.
5. There is currently an urgent need for a refined myth (or set of myths) of telecommunications, which comprehends the 'realistic' elements of the dominant current myths; which incorporates certain basic aspects of current trends and patterns in the global political economy inadequately incorporated within these myths; and which can provide a heuristic guide to telecommunications strategy for the twenty-first century.

The chapter is hence organised as follows. The first section sketches some of the empirical evidence that in the telecommunications sphere itself and in the global political economy since the end of the Second War, we have been undergoing a period of sustained and significant *structural* political-economic change. It also outlines some of the bases for an argument that such periods require different analytical approaches than periods characterised by relative structural continuity.

The next section considers the nature of myths as strategies of uncertainty-reduction, and critically analyses some current telecommunications-related myths.

The final section – the chapter 'Conclusion' – briefly outlines some questions that can aid in assessing our current myths of telecommunications and enhancing our capacity to come to grips more effectively with the radical sea-changes that have conditioned and continue to define the nature of our transition towards a new global informational order.

CONTEXTS OF THE TRANSITION IN THE GLOBAL INFORMATIONAL-ECONOMIC ORDER

Periods of rapid and substantial political-economic transformation entail different conceptual approaches than do periods characterised by relative stability, continuity, and the persistence of social-economic patterns and processes. The latter type of period tends to be theoretically accessible by means of 'mechanical' or algorithmic interpretative frameworks, where the relative constancy or determinacy of the parameters of models of the system being analysed can be more or less taken for granted. The latter type of period, in other words, is susceptible to approaches focused on quantitative changes *within* a given system rather than on *qualitative* changes in the *structure or constitution* of the system itself.

The former type of period, however, directly necessitates a greater concern with modes of analysis that have the capacity to address issues of structural *change*, and that treat elements of structural continuity not as givens, but as phenomena requiring explanation and interpretation. Since an algorithmic theory of structural change is currently (and perhaps in principle[2]) beyond our capacity, periods of transformation require greater emphasis on heuristic, open-system, conceptual frameworks. In the theoretical world of transitions, *mutatis mutandis* (not *ceteris paribus*) rules. Moreover, the political economy of transitions presupposes a dialectical interaction among the elements of the system undergoing transition, entailing processes of mutual and reciprocal causation and determination. These processes involve the interdependence of quantitative change and qualitative system change: quantitative change in the elements of a system produces qualitative or structural change in the system, while these qualitative changes in turn effect quantitative changes in system elements.

The preceding introductory remarks have been expressed in fairly abstract terms. The remainder of this chapter is intended to ground them in concrete examples and more precise definitions. Yet at this point, a preliminary stock-taking is likely appropriate. The foregoing discussion has suggested that periods of rapid and substantial transformation require different modes of analysis from periods characterised by relative continuity and stability of social-economic relations. What makes this distinction particularly significant is that the period extending from 1945 into the twenty-first century appears to be such a period of transition.

This last proposition is potentially open to objections from two basic perspectives. In a strict sense, a serious student of the pre-Socratic philosopher, Heraclitus (who argued that we can never dip our hand into the 'same' river twice, because new waters are ever flowing in upon us), would be entitled to suggest that *all* periods are periods of transition, by definition. Hence, singling out *any* period (such as the period from 1945 to, say, 2010) as a 'transition' period is redundant. Under this rubric, of course, purely algorithmic models should not be relevant under *any* circumstances. In this sense, the Heraclitean approach is the radical limit of the position outlined in the opening sentence of this section.

From a strict methodological standpoint, the pure Heraclitean argument is virtually unassailable, notwithstanding the attempts by Quine and other philosophers to domesticate it. Yet from a pragmatic standpoint, the Heraclitean flux appears to be incapable of dealing with the elements of real or perceived relative *continuity* that exist, at least over some periods, in many social-economic systems. Hence, even if it is correct, it is also incomplete.

The second basic type of objection to the argument that the post-1945 period has been a period of global transition, requiring distinct modes of analysis, comes from the opposite direction. It has two analytically separable components, an empirical one and a theoretical one. Empirically, it is grounded in the notion that even if there have been significant changes, these changes have not *fundamentally* altered the *basic* social-economic framework that has existed throughout the twentieth century.

This notion appears difficult to sustain, although it would be possible to argue that at least some of the changes that have occurred since the Second World War are continuations of trends that had begun by 1900 or even earlier. To begin with some of the more obvious structural shifts: in virtually all of the advanced capitalist nations since 1945 the proportion of women in the wage-labour force has risen rapidly, the sexual division of labour has altered, divorce rates have dramatically increased, the number of single-parent families has risen, birth rates have fallen dramatically, average life expectancy has increased, and the centre of gravity of the age-structure of the population has risen. The early 1990s, in part as a result of the extended global recession, have already produced significant strains on the health and social welfare systems of the advanced capitalist nations, and a continuation of present trends could well produce a global fiscal crisis of the nation-state by or before the second decade of the twenty-first century. Some might argue, with reasonable justification, that this crisis is already in train.

In the global political-economic sphere, the period from 1945 to the 1960s witnessed the virtually complete liquidation of the classic modern capitalist-imperialist colonial empires, and their supplantation by various forms of neo-colonialism. The post-1945 period also witnessed an acceleration in the rate of growth of transnational corporations (TNCs), to the point where they now produce over a fifth of Gross World Product. The growth of TNCs has been associated not only with intensified competition among TNCs, but also with the development of new forms of co-operation and symbiosis among them.

These globalising trends have been associated with a spatial redistribution of material commodity production from advanced capitalist countries to less developed countries and from wealthier to poorer regions within many of the advanced capitalist countries themselves. Correspondingly, the advanced capitalist nations have experienced a substantial intersectoral shift since the latter part of the nineteenth century from the production of material commodities to the production of commoditised services and information. In North America, for example, material commodity production in 1870 constituted about 70 per cent of Gross Domestic Product (GDP), as conventionally measured, with commoditised services constituting about 30 per cent of GDP. By the early 1990s, these percentages had been almost exactly reversed. Moreover, this dramatic quantitative shift was associated with a major qualitative shift, involving a significant commoditisation of goods, services, and information production and distribution activities that had previously occurred outside the ambit of the price system, in households and other institutions not directly susceptible to the logic of the market.[3]

The growing commoditisation of information has had both quantitative and qualitative dimensions. Quantitatively, the commoditised communications sector, as defined in Machlup's pioneering work[4] and augmented by estimates of the output of the transportation sector, currently constitutes over half of conventionally measured GDP and paid employment in North America. Qualitatively, the rise of commoditised communication has had several implications. In the first place, *ceteris paribus*, it has inflated or overstated the rate of growth of GDP as a measure of social output and social welfare, in so far as the commoditisation of formerly non-market activities has lowered the proportion of non-market to market production in the economy as a whole.

Second, the rapid rise of commoditised information has been associated with the rise of a host of new products, processes, and services. At this point, *ceteris paribus* no longer holds. The newer technologies range from television (in its original black-and-white and its colour versions) and FM radio to mainframe, personal, and portable computers and software, telefacsimile (FAX) systems, MODEMs, analog and digital audiocassette tape and compact disc (CD) recorders and players, CD–ROMs, videotapes and VCRs, videogames, 'virtual reality' systems, laser holography, cable networks, fibre-optical systems, telesatellites (direct-broadcast, DBS, and others) and satellite dishes and transponders.

Underlying the development of these interrelated technologies have been significant advances in miniaturisation (from transistors to printed circuits to microchips), digitalisation, signal-compression, multiplex signalling, and signal-acceleration. These advances, singly and in combination, have led to a dramatic relaxation of technical communication capacity limits and to a sharp drop in communication costs, particularly in the sphere of electronic information distribution and processing.

The rapidity and magnitude of these changes have intensified and been intensified by processes of both media-competition and media-symbiosis, at both technical and organisational levels. Corporate struggles centred on competing communication technologies and systems have proliferated in many spheres, in arenas including the marketplace, governments and regulatory agencies, and the courtroom. Numerous examples will come readily to mind: the competition between Betamax and VHS videocassette technology; among Apple, IBM and others in personal computers; among analog audio tapes, compact discs (CDs) and digital audio tapes (DATs) in audio recording (omitting reference to vinyl LPs, 8-track tapes and other dinosaurs); among Nintendo and Sega and their competitors in videogames; and between Microsoft and its competitors over the proprietary rights to the term 'windows'.

Perhaps even more significant than the foregoing struggles are those that have involved competition among and within national and international systems of information-distribution. The struggle among the United States, Japan and Europe over the standards to be established for high-definition television (HDTV) is a case in point. What is at stake here is not only the number of lines that constructs a TV image, but also in part territorial control over the hardware and software production that sustains the world's most important mass medium. The intense battles in North America between cable and direct broadcast satellites (DBS), on the eve of accelerated diffusion of linked interactive television-computer-telephone-FAX-MODEM home-based informational centers, have led to expenditures of millions of dollars on documentation for regulatory agencies, and of billions of dollars on pre-emptive, offensive and defensive, fibre-optical, signal-switching and other system upgrading among the competitors.

The competitive aspect of these changes has likely been more dramatically portrayed in most mass media than the symbiotic aspects, in part because the *agon*, or struggle, is generally more newsworthy, under the present rules of the media game, than the deal. Yet in the long run, present indications of media-symbiosis, the synergistic integration of media at both technical and organisational levels, would seem to be at least as significant as evidences of media competition. The IBM–Apple *rapprochement*, the increase in mergers, acquisitions, and stock transfers between US satellite and cable companies, the growth of transnational consortia in the communications sphere, and the growing pressure for standardisation, inter-system compatibility and the international preservation and extension of private or corporate property rights in information have all contributed to the potential for a more centralised system of informational control, concentrated in private corporate hands.

The foregoing developments have had three further consequences. First, a widening gap has developed between the scientists and technologists of the new global informational-economic order and most conventional social scientists and political economists. Most of the latter clan are not equipped to deal with the *social* ramifications of the new technologies because they do not adequately understand their *technical* capacities, limitations and synergistic potentials. It is equally true that many technocrats begin from their own areas of specialisation and expertise and move outward from these security zones (if they move outward at all), focusing on apparently immediately technically feasible or profitable short-run synergistic/symbiotic possibilities, with potential broader social-economic ramifications of such developments being relegated to a distant third tier of serious analysis.

At the heart of an enormous body of specialised knowledge, for economically intelligible reasons, there has hence arisen an enormous vacuum of organised general ignorance. Bob Dylan's 'Something is happening, and you don't know what it is, do you, Mr. Jones?' could well serve as the emblem of the global informational economy as we approach the twenty-first century. This observation by no means implies a state of stasis or inaction within the global economy in the informational sphere. Quite the contrary: in the kingdom of the blind, those with one eye and sufficient control over wealth to sustain their vision are eminently well situated to capitalise on apparent short- to medium-term opportunities. The principal structural bias of the present situation hence appears to be towards a co-opting of myopic, immediately profit-oriented knowledge in the service of the objectives of private corporate capital, combined with a foreshortening of the global social time-horizon of decision making.

The third major consequence of the foregoing developments has been a major alteration in the global structure of social-economic space and time: in the organisational and material-energy resource spheres, to be sure, but also, and more significantly, in the informational realm. If informational, organisational, and material-energy resources constitute the three primary resources of all economic systems,[5] then the dramatic increase in the potential speed and the lowered cost and increased volume of informational flows, relative to those of organisational and material-energy flows, have significantly altered the global structure of control over time and space.[6]

Related to, but distinct from, the preceding developments, the 1980s and early 1990s witnessed the effective disintegration of the Warsaw Pact, the practical elimination of Soviet hegemony in Eastern Europe, the reunification of East and West Germany, the conversion of the USSR into the CIS, and the re-emergence of the nationalities question within the former Soviet Union and, in even more intense form, within former nation-states such as Yugoslavia. The unexpectedly high costs of German reunification have significantly increased the pressures on the political-economic capacity of the former Federal Republic of Germany and accentuated centrifugal tendencies within the European Economic Community associated with the decline in insecurity incidental to the denaturing of the Warsaw Pact alliance. The preceding events have, for the short term, reversed the secular decline in the global relative power of the United States, without providing either a panacea for its serious underlying problems or a clear-cut system of hegemonic control.

Finally, in what is not intended as an exhaustive catalogue, since 1945 the human capacity to cause irreversible damage to the global ecosystem has significantly increased. Sustained population growth has increased the human demands on the planet's exhaustible and renewable resources, particularly as most less developed countries have entered their demographic transitions. The accelerated growth of environmental pollution processes (many of which are long-lasting and cumulative in their effects), as manifested in the increase in industrial and household effluents, emissions, and waste, and in related phenomena such as acid rain, has been only marginally retarded by so-called 'green' initiatives. Exhaustible resource depletion has continued at an increased rate since the Club of Rome's publication of *The Limits to Growth*, and despite the serious methodological and empirical limits of that study, its concerns have not been eliminated by subsequent technological change, as had been forecast by many of the study's early critics. Moreover, the intervening years have witnessed increased threats to supposedly 'renewable' natural resources, as exemplified by phenomena as disparate as deforestation in the Amazon, the moratorium on cod fishing on Canada's Grand Banks, and sustained permanent losses of whole species of flora and fauna. The period has also witnessed an accelerated rate of depletion of the ozone layer and an increased probability of secular global warming, in relation to human use of chlorofluorocarbons (CFCs), fossil fuels, and other factors.

James MacNeill and his collaborators, in their suggestively titled work *Beyond Interdependence*,[7] have argued that we have moved beyond 'the limits to growth' to 'the growth of limits'; human economic activity in the last half of the twentieth century has in their view accelerated the growth of ecological *limits* to *further* economic expansion along what have become the traditional lines. The term 'interdependence', in this context, has as they see it been sufficiently devalued by past usage that it is no longer adequate to freight the meaning of the globally integrated, planet-wide social-ecological demands posed for humankind *as a whole* by our pattern of development to date.

In the foregoing context, *empirically* based arguments against the notion that we are currently undergoing a period of significant global structural political-economic transition appear difficult to sustain. It might still be argued that even if the global political economy is undergoing significant and rapid structural change, models of the sort generated within contemporary mainstream economics, which incorporate decision-making by optimising social-economic agents under conditions of *uncertainty*, are equipped to analyse such situations.

Despite the undeniable formal elegance and the intuitive appeal of some of the assumptions of these models in simple basic situations, however, the mainstream economic theoretical argument against the necessity of incorporating structural political-economic change directly into the analysis of the complex emerging global informational-economic order rests on tenuous and unrealistic foundations. Practically speaking, its usefulness is vitiated by its neglect of bounded human rationality and of the costs of information production and distribution; its individualistic bias (as manifested in part in its failure to provide a model of a social-economic mechanism for standardisation of the partitioning of future states and actions among individual economic agents); and its algorithmic basis (in so far as in its pure form, it is essentially a straightforward formal extension of the standard deterministic 'perfect-certainty' model of economic equilibrium, buttressed by unrealistic assumptions regarding human creative and information-processing capacity, the costs of information production, transmission and distribution via the market and other media, and the implications of time and space for economic activity).

Yet in practice, human beings *do* make individual and social decisions, in periods of relative continuity *and* in periods of transition. The suspicion arises that the analytical difficulties stemming from the discrepancy between theoretical models of behaviour under uncertainty and the realities of human behaviour stem at least as much from the capacity limits of these models as from the inherent irrationality of human social-economic behaviour. Periods of rapid social-economic transition exacerbate the discrepancy and strengthen the suspicion, in so far as they decrease the empirical basis for the assumptions of structural continuity on which these models implicitly rest. Apparently 'irrational' decisions may in fact be 'rational', once costs of information-generation, information-processing, decision-making and decision-implementation have been incorporated into an analysis of the decision process, even if one arbitrarily excludes observable phenomena such as the 'taste' for novelty, variety, experimentation and (bounded) uncertainty from one's analysis.

The present chapter is grounded on the presupposition that the notion of 'myth' provides an alternative approach to behaviour under uncertainty than that provided by the mainstream economic approach, in periods of both relative continuity and significant system transition. This section has argued that a purely Heraclitean concept of change has practical limits, that there are solid empirical grounds for treating the present epoch as one of significant structural transition, and that present mainstream models of behaviour display theoretical limits in coping analytically with periods of transition. The next section of the paper therefore outlines a theoretical analysis of myth and uncertainty, and considers several dominant current myths of telecommunications.

TELECOMMUNICATIONS, UNCERTAINTY AND MYTH[8]

This section of the chapter may be conceived of as an exploration of myth and uncertainty, conducted through an initial abstract discussion of myths as strategies of uncertainty-reduction, followed by an outline of some contemporary myths of telecommunications and the global political economy.

At the outset, however, it should be acknowledged that certain terms have already been consciously used in the chapter in 'mythic' senses, for practical communication purposes. In particular, three terms warrant attention: 'reality', 'order', and 'equilibrium'. The term 'reality' (along with its cognates, 'realistic' and 'real-world') has already been used in the chapter without qualification, notwithstanding the cautionary note implicit in the introductory epigram from Robin Williams. 'Reality', as Williams suggests, is a value-laden and ambiguous concept. As the old saw goes, for the pessimist the glass is half-empty, for the optimist, it is half full. In addition, the term can have both positive and pejorative connotations: an 'unrealistic' view is unfeasible and utopian, whereas a 'realistic' view is hard-nosed and practical; in contrast, an 'imaginative' or 'creative' vision discerns new possibilities in a situation, whereas a 'realistic' view tends to be more staid, prosaic, unimaginative and conservative, in the pejorative sense of these terms. ('Jones made her fortune by investing in securities. Smith lost his fortune by gambling on the stock market'. Hindsight tends to improve one's assessment of the degree of realism of previous views and actions.)

More radically, all concepts (including concepts of 'reality'), as forms of intellectual fixed capital or means of mental production and communication, involve abstraction. A given set of historical circumstances, structures and processes can generate sharply divergent concepts of 'reality', as a function of the different interpretative processes, abstractions, and privileged presuppositions used in generating them. For this reason, the authorial use of 'reality' and cognate terms (in *this* chapter, as in *all* analyses) should be treated as an implicit invitation to explore the mythological biases and presuppositions of the writer. 'Let us compare mythologies' (Leonard Cohen). Finally, as Foucault has suggested, in his *Madness and Civilisation* and in subsequent works,[9] changing social-economic conditions and the changes in structures of power that accompany them alter social definitions of reality and unreality, sanity and madness. The boundaries of reality are inextricably linked to the capacity to exclude the irrational, as defined principally by those with the power to enforce their definition.

Terms such as 'order' and 'equilibrium' also contain their own biases, in both cases connoting a sense of system persistence, continuity and stability that is dubious in periods of major transition. Some economists have themselves been unsatisfied with these biases: Gerald Helleiner has referred to 'the new international economic *dis*order', and James Tobin, a major contributor to macroeconomic general equilibrium theory, has suggested that it might be more productive to treat the economic process as one of 'perpetual *dis*equilibrium'. While the terms have a sufficiently wide currency that they have been adopted as a shorthand in this chapter, they are intended to be interpreted here as containing their antitheses.

The chapter has argued that since 1945, the global telecommunications sphere has already undergone a number of major transformations – 'improvements' in Innis' phraseology – and many of the technological preconditions for widespread and equally dramatic transformations in telecommunications before the year 2010 are also already in place. Rapid technological and political-economic change, of the sort that has recently been and will continue to be experienced in the telecommunications sphere, tends – as Innis remarked – to make understanding more difficult. The element of paradox in his observation is more apparent than real. It is often assumed, at least in some quarters, that an increase in information flows provides an unalloyed social net benefit, and that the effect of an increase in information flows (of the sort associated with recent telecommunication developments) is typically to reduce uncertainty and increase general social-economic understanding. This assumption, however, rests on an image of the economic role of telecommunications which has always been questionable, and which has become increasingly tenuous as a result of developments in telecommunications since 1945.

These changes in turn have drastically curtailed the capacity of conventional economic theory to analyse, and *a fortiori* to develop policy prescriptions for, the global cultural economy. Some of the difficulties posed for neoclassical economic theory by the quantitative expansion and qualitative or structural transformation of the commoditised cultural-communication sector are both fundamental and obvious: the absence of an adequate quantitative measure of information, and hence of the value per unit of information; the fact that the sharp division between 'producers' and 'consumers' necessary for the application of neoclassical efficiency and optimality (or 'welfare') theorems is inapplicable in the process of production and distribution of knowledge; the fact that over 40 per cent of North American GDP as conventionally measured is devoted to communications activities such as advertising, education, communications media, and research and development, all of which demonstrate the empirical irrelevance of the assumptions regarding perfect certainty, costlessly available information, given technology, and given tastes and preferences on which standard economic welfare-theoretic conclusions rest; and the fact that problems posed by the existence of fixed capital, indivisibilities, overhead costs, unused capacity, joint production, externalities, increasing returns to scale, and decreasing costs (any *one* of which raises serious conceptual and practical problems for the imputation of relative prices within conventional economic theory) are *all* endemic to the cultural-communication sector.

In this context, not only is conventional economic theory seriously flawed as a means of developing policy prescriptions for the cultural-economic sector itself; it is also, practically speaking, given the growing quantitative importance of the cultural sector, becoming increasingly irrelevant (and therefore potentially dangerous) as a guide in the determination of *overall* economic policy.

In so far as the telecommunications sector itself is concerned, the current conventional definition of the sector exacerbates these problems. Viewed etymologically, 'telecommunication' is simply *long-distance* communication. Broadcast media, and for that matter printed media and *all* communications media, with the partial exception of direct, face-to-face oral and more intimate dialogue, therefore to some degree constitute forms of telecommunication in this broader sense. From this etymologically based definitional standpoint, then, the hiving-off of certain components of the broader telecommunications sector to create a telecommunications sector for regulatory-administrative purposes can be seen as at best a matter of administrative convenience, rather than as one of inherent logical or theoretical necessity. This 'hiving-off', however, restricts the capacity to see the cultural economy as a whole, and hence to develop an integrated, comprehensive strategy for the entire cultural-economic sector.

There is obviously a need for *some* division of labour in the analysis, policy-determination, and regulation of the cultural economy. No one person can be expected to have *complete* mastery of the internal technological and economic workings of all of, say, the cable industry, the educational system, the fine arts, the independent film and television production subsector, the computer industry, radio-TV broadcasting, and the telesatellite subsector. Yet there is also a need for a better balance between such specialised and detailed subsectoral knowledge and a more comprehensive and integrated vision of the complex set of interrelationships and interdependences among the various subsectors of the cultural economy.

To take the newspaper industry as just one (deliberately off-centre) example, this print medium simply could not function in its present form without reliance on a communications system involving interlinked computers, telesatellites, telephone, telefacsimile (FAX) and MODEM services, and printing systems. This combination is as essential for the *Winnipeg Free Press* or the *Times-Picayune* as it is for the London *Times*, the *New York Times*, or *USA Today*. Given the current definition of the telecommunications sector, however, this set of elements often tends to be treated purely in terms of isolated external demands on the telecommunications sector narrowly defined, rather than as an integral and systematically related part of the telecommunications sector under a broader definition. The net result of such an approach hence tends to be to restrict the social capacity to develop an integrated vision and strategy for the cultural-economic sector as a whole, as manifested in the failure to take adequate account of intra-sectoral interdependences, externalities and synergistic possibilities; consequently incompatible or contradictory specific policies, designed largely in response to the short-run demands of interest groups within particular subsectors; and attendant efficiency losses and increases in uncertainty.

The global development of telecommunications since 1945 has likely contributed, on balance, to increased productivity growth, to an acceleration and reduction in the cost of information flows, *and* to an increase in global uncertainty and insecurity. The reasons for this apparently contradictory set of effects are relatively straightforward. *Ceteris paribus*, the 'improvement' of communications *within a given economic system increases* knowledge concerning the system and its environment, thereby enhances decision-making capacity, and hence increases efficiency and productivity.

This logical chain constitutes the core of the standard neoclassical economic analysis of information as a commodity like any other commodity. The principal limits of the neoclassical-economic analysis of information stem from the inadequacy of these assumptions when *ceteris paribus* no longer holds. In the first place, changes in the cost and character of information flows are a major cause of structural economic change; new rules of the game; the systematic devaluation and revaluation of pre-existing stocks of knowledge; and consequent *increases* in uncertainty. Secondly, it is (although theoretically not impossible) historically extremely rare that a particular communications innovation affects all subsectors within a given communication network or system to an equal degree, and hence the typical effect of a particular communications innovation is to cause problems of unused capacity in some parts of the system, and severe pressures on capacity (bottlenecks and 'information overload') in other parts of the system, of the sort that heighten pressures for technological and organisational change (to re-establish a balance among subsectoral capacities) and correspondingly increase uncertainty. These two factors, as well as alterations in the environment attributable to the effects of communications system change, are often also associated with shifts in the social distribution of effective access to information, on class, regional, sexual, ethnic, linguistic, or large-scale (government or private corporate) *versus* small-scale entrepreneurial lines. Such social-economic information-distributional shifts can alter the potential for the social monopolisation of knowledge, wealth and power and hence contribute to increased uncertainty, insecurity and instability within a given political-economic system. For the foregoing reasons, telecommunications advances have contributed both to an increase in productivity and to a high overall level of uncertainty.

Periods of high uncertainty tend to generate an increase in the creation and dissemination of *myths*, as a strategy of uncertainty reduction. Pure uncertainty arises more pervasively in situations where the old rules of the game no longer appear to hold, where levels of insecurity (and the associated levels of individual and social stress and contradiction) are high, and where the reduced effectiveness of existing social-homeostatic processes has increased the degree of instability within a social-economic system.[10]

Social myths provide means of economising on perceptual, conceptual and communication time, and thus provide increased individual and social control over time. 'Myth' is not a synonym for untruth. On the contrary, historically the most powerful myths owe their temporal longevity and spatial extent to their capacity to embody significant and strategic aspects of reality that in some sense transcend the limits of a particular time and place. This is not to say that myths are abstract. Myths are fundamentally concretely grounded *stories*, about people (or about anthropomorphised gods, animals and plants), and their universality emerges in the context of the particular stories through which they are told. Myths and models, in their capacities and limits, are in principle isomorphic, although the procedures for using them and 'testing' them are often quite different.

The other side of the coin is that no myth is *completely* true: like all models or stories, myths involve an anterior determination and a corresponding selection of which facts or details are most salient or significant (what Joseph Schumpeter, with regard to economic models, called a 'vision' of reality), and hence all myths are characterised by biases or capacity limits. Moreover, the dissemination or distribution of social myths is not costless, and hence the existing distribution of wealth and power tends to determine which myths have the greatest spatial and temporal diffusion, independently of their 'truth'-value. Hitler's 'Big-Lie' theory – that if you tell enough people a big enough lie, often enough, they will come to believe it, or accept it as 'truth' – constitutes the basis of a fundamental principle of much effective propaganda and advertising. Success in altering social perceptions of 'reality' in this context thus depends on access to sufficient material-energy resources and power to enable sufficient repetitions of a myth that its 'truth' appears self-evident.

The power of myths rests as much on their control of unconscious preconceptions as on their logic at the conscious level. Hopes and fears, desire and revulsion, as well as aesthetic aspects such as the beauty and symmetry or ugliness and disproportion of a particular myth and the extent to which a particular myth generates social-psychological resonances in its audience, or 'strikes a responsive chord', given a pre-existing mythic framework, are among the forces (apart from the factors noted above) that determine the ultimate durability and extent of particular myths.

One of the most powerful telecommunication-based myths regarding the current global political economy is still that promoted by Marshall McLuhan,[11] which centres on the image of a 'global village', and which assumes in part that the 'linear', bureaucratic (or 'centre-margin') structures erected on a foundation of print and literacy have been undercut (and will ultimately be overturned) by the penetration of electronic media into the global economy. This myth presupposes that electronic media have created a 'tactile' world of multiple 'centres without margins', in which humans have been 're-tribalised' and converted into 'hunters and gatherers' of information, rather than of material goods. Telecommunications, in McLuhan's vision of current 'reality', are the technological agents of a global Chardinian epiphany in which electronic media annihilate space and time.

This sketch is to some extent a parody of McLuhan's myth in its more complex forms, but as a representation of the version of the myth that has received wide currency, it does not involve significant reductionism. The ideological strengths of this myth of the regaining of identity are obvious: it provides an image of hope centred on the further development of electronic telecommunications media that are already in place, which will eventually result in the harmonious global dance of multiple autonomous and independent centres of knowledge, wealth and power. Moreover, the myth has no explicit programmatic content: the millennium of the global village (in the crudest versions of McLuhanism) is an inevitable result of the continued development of technological trends that are already apparent, and the primary social task is to *understand* these trends and their social effects, not to attempt to transform society in the light of 'rear-view mirror' thinking that will only prolong and render more painful this period of transition. Present uncertainties, in the context of this myth, are largely failures of imagination.

The attractiveness of this myth is considerable. It has a populist aspect; it requires no action; it accounts for a number of current trends; it provides an image of the regaining of social identity and harmony; and it explains current uncertainties as a transitional phenomenon. The limits of the myth, however, are equally considerable. It has no effective theory of the determinants of political-economic power; it hence involves (at least in its more vulgar forms) a reductionistic sort of technological determinism, in which mechanisms and processes of determination are unspecified or underspecified; and (perhaps most seriously) its teleological bias inhibits recognition and analysis of factors that may retard *or* promote, the progress towards the millennium.

A number of variants of this 'global-integration' myth have been developed as means of uncertainty-reduction, all of which bear on the question of the appropriate telecommunications strategy for the twenty-first century. Some of these myths relate to the telecommunications sphere itself; some relate to the overall environment of telecommunications. One of the most powerful and inclusive of these myths, notwithstanding its major inadequacies as a policy guide (for reasons sketched above), is the 'free-market' myth. This myth draws on the image of the costless and perfectly functioning market incorporated in the axioms of basic neoclassical economic theory, and emphasises the role of *competition* among profit-maximising firms in generating an 'efficient' allocation of resources in the short run, and rapid technological change and productivity growth in the longer run.

In its neoconservative variants, this myth also tends to rely on an image of government as a vast, incomprehensible, slow-moving, unwieldy, and monolithic bureaucracy that rewards incompetence, inaction, empire-building and obfuscation. The myth, furthermore, ignores or downplays the numerous situations, acknowledged within more sophisticated neoclassical theory, in which 'market failure' can occur and government regulation and/or control becomes necessary to ensure the socially efficient allocation of resources. 'Market failure' will typically occur when any one of the following phenomena exists: indivisibilities; increasing returns; joint production; resource-consuming or incomplete sets of markets; transactions costs; interdependent utilities; externalities; imperfect competition, including so-called 'natural monopolies'; public goods; common-pool problems; spatial competition; or divergences between private and social discount rates. Ironically, this catalogue of the situations in which market failure occurs is at the same time a reasonably precise description of the principal economic characteristics of the telecommunications sector and of the cultural economy as a whole.

Yet the 'free-market' myth – which provides the ideological basis for such government policies as deregulation, privatisation, the so-called Canada–US and North American Free Trade Agreements, the treatment of information as private property rather than as a public good, and the related shift from 'universal-access' to 'pay-as-you-go' principles in the information market – undeniably has considerable political-economic power. Much of the force of the myth stems from its simplicity (cynics might say its simple-mindedness): 'Just let the free market work, and the cornucopia is yours'. As with the McLuhan myth, no demands for personal action or responsibility are imposed on one who subscribes to it. Moreover, the image of 'big government' as an impersonal, arbitrary and inefficient monolith, apart from its populist, 'anti-bigness' appeal, almost invariably strikes a responsive chord in anyone who has had to deal on a matter of some personal importance with one of the rude, obtuse, rule-bound, inflexible and unhelpful petty tyrants that tend to infest the lower ranks of any large-scale bureaucratic organisation, public or private, municipal, provincial or federal. (Most individuals who have had, say, to make a claim on an insurance policy with any private-sector insurance company, or to cross swords with a Book-of-the-Month Club's computer, will likely have similar resonances, but the power of the 'free market' myth is such that its bias, or filter-system, tends to put a higher weight on experiences of the former type, and a lower weight on the latter, so that the net result is a biased, but existentially grounded, prejudice against 'Big Government', and a corresponding prejudice in favor of 'free markets'.)

The limits of the 'free-market' myth, however, are quite profound, and nowhere more so than in relation to the telecommunications subsector and the cultural economy as a whole. The 'free market' functions 'efficiently', from a social standpoint, only under a highly restrictive set of assumptions, virtually *none* of which is wholly satisfied in practice within the cultural-economic sector, as has been indicated in some detail above.

In part as a result of the significant structural changes in the former Warsaw Pact countries, myths concerning the *imminent* collapse of capitalism under the weight of its own contradictions and its supplantation by international socialism have tended to be shelved for the time being. Socialist students of the global political economy[12] have tended to focus their energies on critical analysis of contemporary trends rather than on the articulation of myths of transition towards socialism. Yet there are still suggestions that the global development of telecommunications and TNCs is finally creating the basis for the establishment of Marx's classical myth of the transition to socialism: the creation of a full-fledged integrated world market; the concentration and centralisation of capital; the commoditisation and socialisation by private capital of global production; and the intensification of the contradictions of capitalism itself, those foreseen by Marx as well as those posed by subsequent technological, cultural and global ecological developments.

In this myth, the socialist experiments of the Soviet Union and Eastern Europe were simply premature, either because the material basis for socialism on a world scale was inadequate in 1917 or because of incapacities and contradictions within the Soviet Union that led to its dismantling. Subtler variants of this myth also emphasise both the more positive aspects of the accumulated knowledge and practice of socialism within COMECON, and the fact that the transition from a relatively centralised to a more decentralised system entails a major investment of time, energy and resources in the creation of new capacities and institutions, substantial transition costs, and significant uncertainty regarding the actual transition path and the criteria for and likelihood of a successful transition.

In addition to these global myths, some other more specific myths have developed:

1. The nation-state is now obsolete, as a result of telecommunications developments and the development of the transnational corporation, which have produced a genuinely global market that transcends national boundaries, and renders the nation-state currently irrelevant.

2. The world is now, of necessity, becoming a world of trading blocs; individual nations, particularly smaller nations, can no longer survive on their own, but need to become part of larger political-economic entities.
3. Socialism (and/or traditional capitalism) is now obsolete; the Cold War is over; and we are now witnessing a process of convergence of political-economic systems.
4. 'Bigger is better': in order to compete successfully in the world market, nations need to eliminate barriers to the centralisation of capital domestically, so that their firms have the capacity to compete internationally.

Over and against these myths, another set of myths, with a different implicit political-economic agenda, has also developed.

5. The nation-state is not yet obsolete, and (particularly for nations at the margins of the global economy) efforts to control foreign political, military, economic and cultural domination are an essential component of strategies for transition, both domestically and to promote a more firmly grounded international political economy.
6. Neither capitalism nor socialism is yet dead; the present transition period is one characterised by significant uncertainty and contradictory tendencies; and renewed talk of 'convergence' is still decidedly premature.
7. 'Small is beautiful': the world is on the verge of major environmental eco-disaster, and the only solution is a radical alteration in global political-economic policies and practices.
8. A 'New International Economic Order' is required, to deal with inequalities in the distribution of global income and wealth, and to cope with the currently effectively insoluble problems posed by the magnitude of Third World debt.

It is important to recognise that none of the foregoing myths has been rigorously established as 'true' on theoretical or empirical grounds, although efforts at theoretical proof and/or empirical verification have been made in the case of some of them, generally without a conclusive outcome. This very inconclusiveness is what *should* be expected in a period of uncertainty, at the level of myth. To suggest, therefore, that these myths are useless or ineffectual would be to miss the whole point about myths as essential elements in strategies of uncertainty-reduction. *The central point is that myths provide means of acting in the absence of perfect information, and that they focus inquiry in directed ways, which tend to be biased in favour of the interests of those by whom they are promulgated.* In situations of high uncertainty, myths are essential stop-gap measures, to enable both short-run thought and action. In so far as future telecommunications strategy is concerned, however, the crucial question is whether we can build a better myth-trap – or, alternatively put, whether we can develop a framework of analysis that is better equipped to address the practical questions associated with future telecommunication strategy than the frameworks underlying existing myths.

Existing myths will continue to have effective power until they are supplanted by myths or models that can reduce uncertainty to a greater extent. The conviction of the present chapter is that improved models or mythologies *are* in fact possible, and that such models will involve a significant reorientation of cultural and telecommunications strategy. The concluding section of this chapter outlines some elements that need to be considered in such myths.

CONCLUSION

Developments in the telecommunications sphere have been of fundamental importance in determining the patterns of development in the current political-economic transition, and telecommunications satellites have been among the most dramatic innovations in telecommunications. Moreover, recent changes in telesatellite technology and in other media have significantly relaxed former communication capacity limits and sharply lowered the costs of communication via telesatellite.

Uncertainty regarding the implications of these changes has led to the emergence of a number of telecommunication-related myths, of which two are of particular significance. The myth of the 'Information Society' is one. In its more utopian forms, it projects an inclusive, integrated, information-intensive, high-technology global economy in which virtually everyone has effectively instantaneous access to all the information he or she desires, and to all people with whom he or she wishes to communicate. This utopia is a world of potentially flexible work hours and locations, with production processes integrated by communication networks utilising computer-telephone-telesatellite links among home-based information-entertainment systems. It is characterised by multiple loci of decentralised knowledge and power, and success in it will require the capacity on the part of individuals to use the available systems creatively. Unfortunately, no one has yet provided adequate estimates regarding how much it will cost to get there from here, nor a programme for the transition.

An alternative myth assumes that present trends are part of a transition (for the medium-term future) towards more powerful and integrated monopolies of knowledge, wealth, and organised force, centred on transnational capitalist corporations, private property in information, and the states of the most powerful advanced capitalist nations. In this vision, the abortive US 'Star Wars' (or Strategic Defense Initiative) programme is being supplanted by what could be described as 'Star Wares': the use of DBS and other satellites to beam potentially hundreds of channels of commercialised and commoditised culture down to waiting satellite dishes, transponders, and fibre-optics-based telephone and/or cable systems.

The system would operate across national boundaries, rendering parochial notions of 'cultural sovereignty' obsolete. It would be funded largely by advertising, subscriptions (pay-TV and its more generalised, text-data-base, cousins), and corporate private-channel rentals. The US, as the world's largest commoditised entertainment producer, along with other major producers of electronic and computer hardware and software, would dominate this system of high-tech global cultural-economic imperialism. The substantial fixed costs of establishing information-processing and analytical capacities to utilise the specialised knowledge available through the system would intensify the monopolisation of knowledge by private corporate capital and accelerate the centralisation of capital. In this context, the world would be increasingly divided between informational 'haves' and informational 'have-nots', both between the advanced capitalist nations and the rest of the world and within the advanced capitalist nations themselves.

At present, the second myth appears to have somewhat greater plausibility, but it is not an inevitability. In assessing the degree of 'realism' of alternative myths of telecommunications, the following checklist is therefore proposed:

1. Does the myth in question adequately comprehend the full implications of global environmental-ecological challenges and of the unequal global distribution of wealth for the vision of the future it contains?
2. Does it display a clear recognition of issues related to costs and timing of the transition? (Costs and timing questions are intimately related, inasmuch as the magnitude of fixed-investment costs, investment lags, the relative power of competing vested interests, and the speed of diffusion of innovations are all interdependent.)
3. Does it display a clear understanding of issues related to the treatment of information as a public good as opposed to as a form of private property?

4. Does it focus on the need for 'high-tech' solutions to current difficulties, or is it alert to the possibilities of what could be described as 'high-culture' or 'culture-intensive' strategies? 'High-tech' strategies often tend to be capital-intensive in character, and to emphasise increasing the productivity of those employed by replacing people with machines, at the cost of increased unemployment. 'High-culture' strategies, in contrast, include both 'high-tech' strategies and activities (such as preschool day care) that are labour-intensive in their inputs and knowledge-intensive in their outputs, and that hence increase not solely productive capacity, but also the capacity to increase capacity.
5. Finally, does it display an awareness of the need for balance between cultural production and cultural distribution? Telesatellites are a medium of information-transmission or -distribution, albeit one with distinct capacities and certain advantages over alternative, terrestrially based media. Moreover, their basic economics, under a number of alternative regimes, appear to be quite healthy. The same is true of most of the present systems of telecommunication and cultural distribution. It is less true, however, particularly in countries marginal to the principal advanced capitalist nations, of the sphere of cultural *production*. A litmus test of any myth of telecommunications is therefore the thoroughness with which it also envisions what culture is produced and how and by whom culture is produced, and the extent to which the myth incorporates a vision of the appropriate balance between cultural production and cultural distribution.

The present chapter has been intended primarily to raise questions, rather than to provide definite answers about the future development of telecommunications in this period of global transition. If it has been successful, however, it should have provided an image of some current myths of the transition and their limitations, an outline of the global political-economic context within which the telecommunications sector will develop, and an indication of the sorts of questions that need to be raised in order to establish the components of these myths that are of most use in developing strategies for the transition. It should also have suggested that conventional economic tools, while they do have their uses in analysing some aspects of the transition, cannot be relied on to provide a full analysis because of their limits in analysing both communications phenomena and situations of structural change. In periods such as the present one, critical comparative study of the myths generated by the transition can offer an alternative avenue of insight into the changes themselves, the hopes and fears that they engender, the configurations of knowledge, wealth, and power that can potentially emerge at future stages, and the most promising spheres for strategic intervention.

Notes

1. Harold Innis, *The Bias of Communication* (Toronto: University of Toronto Press, 1964) p.31.
2. Cf. Roger Penrose, *The Emperor's New Mind* (New York: Vintage, 1990) on the non-algorithmic character of the creative process.
3. Cf. H.A. Innis, *Essays in Canadian Economic History* (Toronto: University of Toronto Press, 1956) pp.252–72; and Ian Parker, 'Commodities as Sign-Systems', in Robert Babe (ed.), *Information and Communication in Economics* (Boston: Kluwer, 1993).
4. Cf. Fritz Machlup, *The Production and Distribution of Knowledge in the United States* (Princeton: Princeton University Press, 1962).
5. In conventional economic theory, two schemas for defining primary economic system resources have become standard. The first is that centred on the classical economists' land, labour, and capital, with 'entrepreneurship' occasionally thrust in as a fourth 'factor', in honour of the work of Frank Knight, despite its incommensurability with the other three 'factors of production'. The second, the backbone of contemporary mathematical economics, typically allows for a vector of non-produced, heterogeneous primary inputs (often including homogeneous labour or a set of types of labour, and even in some cases including a rather ill-defined measure of 'capital', whatever that is supposed to represent in such models), but makes no direct allowance for the role of information or organisation in production. The treatment of informational, organisational and material-energy resources as *primary* economic system resources in the present chapter is not intended to deny the problems of measurement that this usage entails. Nonetheless, as this footnote suggests, the limits of the alternatives available in conventional economic theory imply, at a minimum, that for heuristic purposes in periods of rapid structural change, this alternative definition has greater flexibility and capacity, and moreover is, at least to some extent, susceptible to quantitative measurement.
6. In this context, time and space refer both to duration and territorial extent, respectively, and to the human constructs and processes that sustain our systems and concepts of time and space. The concept of control over time and space was developed initially, in much more complex form than that sketched here, by Harold Innis in works such as *The Bias of Communication* and *Empire and Communications* (Toronto: University of Toronto Press, 1972).
7. James MacNeill, Pieter Winsemius and Taizo Yakushiji, *Beyond Interdependence* (New York: Oxford, 1992).
8. This section of the chapter contains a revised and expanded version of an analysis contained in Ian Parker, 'Myths and Realities of Telecommunications for the Twenty-First Century', *Canadian Journal of Communication* (Spring, 1990) pp.33–45.

9. Cf. *inter alia* Michel Foucault, *Madness and Civilization* (New York: Vintage, 1973) and *The Order of Things* (New York: Vintage, 1973).
10. This sketch draws on W.T. Easterbrook's concept of the 'macro-uncertainty' environment of social-economic systems, which includes three dimensions of macro-uncertainty: *pure uncertainty*, related to the impossibility of precisely knowing the future; *insecurity*, related to threats to the continued existence of an economic system over space and time; and *instability*, emerging from inadequate adaptation to altered structural relations within a system or between the system and its environment. See W.T. Easterbrook, *North American Patterns of Growth and Development* (Toronto: University of Toronto Press, 1990) pp.xvi–xxvii, 3–20, and the articles by Easterbrook cited on pp.i–ii.
11. Marshall McLuhan, *Understanding Media* (New York: New American Library, 1964).
12. For a useful stocktaking, see Ralph Miliband and Leo Panitch (eds), *Socialist Register 1992: New World Order?* (London: Merlin, 1992).

4 Surveillance and the Global Political Economy

Martin Hewson

Surveillance, it seems, is everywhere. There is the constant monitoring of the local variations in global markets by producers of consumer goods and services. There is the accelerated management of information that enables multinational corporations to engage in global systems of flexible specialisation. There is the expanding monitoring of debt and creditworthiness by bodies such as the International Monetary Fund. There is the intelligence produced by orbiting satellites that intensifies the information density of contemporary war. It can appear that an insidious power is spreading and penetrating all places to create a totally monitored world.

But surveillance can also seem uninfluential or secondary. It has only a virtual existence compared to the apparently solid agencies, institutions or corporations to which it is linked. From this point of view surveillance becomes a useful tool, at best, of social entities such as states whose interests and actions have their bases elsewhere. Alternatively surveillance might be understood as part of a level or dimension of some wider structure such as the global political economy that is constituted by other principles. If understood as instrumental or partial, then, mere monitoring does not seem very powerful.

This chapter aims to specify the role of surveillance in relation to the global political economy without either elevating it into an omnipresent power or denigrating it as a secondary phenomenon. Surveillance power arises from the accumulation of information and this exercises a particular kind of influence in modern globalising society. The first section of the chapter identifies this form of collective power and its sources in some of the most basic practices of modern communication and expertise. The second section provides an overview of the two leading approaches to the study of the global political economy in order to explain their neglect and misreading of surveillance. The final section specifies the significance of surveillance in the global order in terms of the types of modern social forms that it enables. Drawing especially on the late work of Michel

Foucault and on Anthony Giddens, the intention behind the chapter is to propose a range of concepts by which global surveillance might be better interpreted. This contribution is suggestive rather than conclusive; ironically, firm conclusions add more to surveillance itself than to our understanding of it.

SURVEILLANCE POWER AND ITS SOURCES

The role of the global information order in relation to the global political economy may be understood in several ways. Treated in association with world-wide market institutions, information may be analysed as a commodity, exchanged in ever-more globalised circuits. Alternatively, understood in terms of the concept of belief system or ideology, its role may be seen as part of a global hegemony. This chapter takes a different approach to suggest that information helps constitute a source of global surveillance power. Simply defined, surveillance is the social power engendered by the accumulation of information, where accumulation derives from the intersection of communication media and expert systems. If information is a power resource, experts are its brokers. The present chapter thus brackets the commodified and ideological aspects of global communications to focus instead on the power of surveillance.

The exercise of power has two distinctive modes. One is a matter of facilitation, enablement and creation, the other works as control, exclusion and prevention. Most theorists of the global political economy attend to the latter mode. In debates over hegemonic power 'three faces' appear: the structures that prevent subordinate classes and other groups from achieving emancipation; the institutions that exclude peripheral states from full participation; or the actors, usually dominant states, that control the behaviour of others.[1] By contrast, surveillance power is primarily facilitative in operation.

To illustrate the exercise of surveillance, consider what was involved in the establishment of the first modern statistical office, the Inspector-General of Imports and Exports, in London at the end of the seventeenth century.[2] The office itself constituted the centre of a monitoring system that served to gather the records generated by the customs service into a data bank, from which could be assembled a 'balance of trade'. This new informational resource helped to enable many modern practices. It shaped the emergence of modern nations, for the viewer of a balance of trade is more likely to conceive a distinctive national population and national wealth in place of the feudal hierarchies. Balance of trade records also affected the practice of governing. With surplus in the 'balance' now a concrete measure of policy success, policy itself came to rely on measurement directed at a national population and wealth. These are the kinds of development enabled by surveillance power.

The same example illustrates the sources of surveillance. First, it entailed the *visualisation* of a domain with a virtual existence on a set of records – in this case an entity that possesses a 'balance'. Second, it relied on the *mobilisation* of information, as customs agents listed in time-series the ships calling at port, classified the goods, copied, filed and communicated the concentrated data to the centre. Third, the example illustrates the importance of the *informability* of practices, for it is only the relatively transparent or observable activities of open trade rather than smuggling that yield useful quantities of information. Fourth, linking together the constituents of surveillance is *expertise* carried by specialised agents (such as customs officers) organised in expert systems. The remainder of this section discusses at greater length these general features of the accumulation of information into surveillance power.

The place where information becomes visible is most often the surface of a printed inscription or electronic screen. Print and screen media are similar in respect of their two-dimensional interface with the viewer. That flat plane is essential to surveillance because on it domains are seen, and only visible domains can be examined, compared and manipulated. A key practical problem in surveillance is thus how to transfer convoluted reality on to flat surfaces, or as Edward Tufte puts it, how to 'envision information' on 'flatland'.[3] The problem is how to enhance the clarity, dimensionality and density of representations (such as maps, tables, diagrams and so on) without destroying their intelligibility with clutter and confusion. Tufte identifies a range of techniques. One is to create arrays of multiple items of the same form, which makes visible both individual differences as well as a comprehensible narrative of the domain as a whole. Another is the separation and layering of data, most notably by tabulation, which makes visible a uniform framework and its internal variations. A final technique reveals an overall pattern by enhancing detail, in which both the overview and the minute details of a domain become visible. Each technique is a variation on the theme of creating together the individual case and the total population, the variant and the average, the singular and the panoramic. These are paradigmatic modes in which domains (such as a global economy or a global polity) become intelligible to the monitors who look at flat surfaces.

Visual representations must be mobilised in order to become powerful as surveillance. Modern society is coeval with the use of printed inscription and electronic transmission as leading vehicles of mobility. Both have a bias toward enabling the accumulation of information by contrast to pre-modern oral or scribal media.[4] Print culture produces a continuous 'knowledge explosion' on inscriptions that are both numerous and immutable, thus able to be circulated, gathered and stored without loss or corruption.[5] The bias toward accumulation is actualised in conjunction with expert systems. Bruno Latour introduces the concept of 'cascading' to refer to techniques for making visualisations combinable, readable, reproducible, and of modifiable scale. The file is a classic technique of cascading information. In files, obscure areas become clear, distant domains become near, whole populations become accessible. The cascading of information gathered from widely spaced areas thus enables the globalised co-ordination of practices. Latour uses the term 'centre of calculation' to describe the containers in which cascading is produced. Such places give rise to emergent properties, new resources separate from the originally observed domains.[6] It is these emergent resources (such as the visibility of standardisation and differentiation in a population) that help to constitute surveillance power.

In addition to envisioning and cascading, surveillance depends on the informability of domains. In the example of the fabrication of the balance of trade, there was a relatively high degree of disclosure, for the movement of ships in ports could be continuously observed. It would be misleading to conceive informability as an imposition. Everyday use of printed material, for instance, has as a concomitant a high availability to disclosure. An important source for the unprecedented informability of modern society is, to repeat Elizabeth Eisenstein's phrases, the 'knowledge explosion' of 'print culture'. The past becomes informable as 'history' with the circulation of permanent records of bygone times. Distant and distinct places become informable with the extraction and concentration of diverse records from them.[7] As a result, modern society is an informable society, both on an intensive and an extensive scale, as the extent of disclosure to surveillance expands across the globe.

All the sources described above involve expertise, a final constituent of surveillance. Expertise is associated with specialised knowledge. Discursive literacy is one example, indeed it may be said that literate persons are the basic agents of surveillance, monitoring themselves, gathering and yielding information about themselves. Again, this is to emphasise that surveillance arises not from the domination of an already-powerful centre. Rather, it arises from collective practices that facilitate monitoring. Quintessential techniques of surveillance such as the population census or market research depend for their effectiveness on voluntary and literate subjects. Other kinds of expertise are more specialised and more professionalised. The 'career' of the professional is itself an exercise in self-surveillance in that information is accumulated in the form of appointment diaries, career plans and examinations that organise the time of training and practice into segmented and stadial units of development.[8] Within the space of a career and a distinctive office the professional frequently has a good deal of autonomy both from superiors and clients. The sources of surveillance circulate through networks of associated and organised experts. Such expert systems are able to extend themselves across the globe.

The sources of surveillance, then, are collective and plural. They penetrate deeply into modern globalised life. The remainder of this chapter turns to consider what surveillance power sources help to create. Too often surveillance is treated as if it creates nothing at all, or as if it creates an all-encompassing global dominance. The following section discusses the implications of surveillance in relation to the leading approaches to theorising the global political economy.

THE GLOBAL POLITICAL ECONOMY: INSTITUTIONALIST AND HOLIST APPROACHES

In the study of international political economy there is a fundamental divide. In addition to the methodological division between rationalist and interpretive perspectives,[9] there is a substantive difference between holist and institutionalist approaches. For holists, the global political economy is an integrated totality. They generally conceive globalisation as a process of incorporation into the dominant global order. For institutionalists, the global political economy is composed of separate world market and political institutions. Globalisation is driven by the restlessness of the market. Neither approach accounts for surveillance. Because surveillance has plural sources, it tends not to have the effect of incorporating social forces into the dominant global totality. Because it is a collective and facilitative power, surveillance has a marginal role as an instrument of pre-existing market or political institutions.

World markets and political institutions are poised in external relation to one another according to the institutionalist approach of both neo-liberals and neo-realists. To start with the market, its chief attribute is dynamic expansiveness. As far back as 1711, Joseph Addison gave a glowing description of global market dynamism. Like today's observers, Addison remarked on the importance of bargaining: 'Factors in the Trading World are what Ambassadors are in the Politick World; they negotiate Affairs, conclude Treaties'. Also like many today, he thought market contacts 'maintain a good Correspondence between those wealthy Societies of Men that are divided from one another by Seas and Oceans', by a 'mutual Intercourse and Traffick among Mankind'. Beyond simple interconnection, Addison articulated the concept of interdependence: 'a kind of Dependence upon one another ... united together by their Common Interest'.[10] Put more formally, markets globalise for three reasons. They offer opportunities for forms of human action that are rational in the sense of being instrumentally efficient at deploying available means to achieve desired goals. Second, they enhance the benefits of specialisation and the division of labour, which is the basis of interdependence. Finally, expanding markets have been seen consistently as a 'civilising' force, promoting a diversified way of life in a pacific civil society.[11]

A problem of governance arises when market dynamism propels itself across the globe, apparently beyond the jurisdiction of any single authority. There follows an unending contradiction. While the task of regulation is essential to the existence and expansion of markets, it is continually outflanked as markets keep expanding beyond the regulator's ambit. The debate on this has long counterposed nation-states against international institutions and their respective champions' realists against liberal internationalists. The debate turns on the convention that the institutions that count are the ones that are authoritative. The debaters thus concern themselves with discovering the exact location of authority. One side supposes that the sovereign state monopolises authority in the form of legitimate violence and the other claims that international institutions also possess legitimate power (or that it should be redistributed to them) as a result of the material benefits that derive from their regulation and management.[12] Either way, the basic characteristic of institutionalism is to seek to understand the relation between market dynamism and the global distribution of political authority.

There is little space in the debate for surveillance. If it makes an appearance it is as an instrument of a political authority, a means for more efficient regulation of a portion of the market. Robert Keohane, for example, discusses the role of 'information-producing "technology" that becomes embedded in a particular international regime' as a means of monitoring compliance with the rules of the regime. Keohane's hypothesis is that the more effective a regime is in providing information about other actors' intentions, resources and negotiating positions, the greater the demand for and persistence of international regimes.[13] The institution has conceptual priority and surveillance is a secondary matter.

Holists believe that institutional separation is illusory because the global political economy is a whole that subsumes different dimensions within its compass. Aspirations of a holist kind to apprehend the entirety in all its dimensions can be found in world-system theory or in the neo-Gramscian approach. The remainder of this section asks what happens when surveillance is brought within the incorporative field of a holistic notion of the global political economy.

Immanuel Wallerstein's 'world-system' perspective is well known. It posits a global totality within which the ensemble of states, movements, households and so on have their place in a hierarchy of core and peripheral regions. What constitutes the whole is the world-wide division of labour and its circuits of market exchange that incorporate all other kinds of social relations within its unitary logic.[14] If surveillance is added to the world-system, it becomes a functional device for ensuring the continued cohesion of the whole. Attributing to states a purely superstructural location is a feature of world-system theory that Bruce Andrews, for example, suggests might be refined by investigating how surveillance power aids 'the constitution and normalisation of states within this structural whole'. Equating surveillance with a subtle and internal imposition of discipline, and drawing an analogy between discipline for individuals and for nation-states, Andrews believes it to be a (highly functional) means by which a national society can be 'incorporated or normalised more cheaply' through the 'socialisation, or training, or disciplining, or normalisation of the body politic'. The result of awarding a fundamental priority to the global whole seems inevitable. Since the world-system cunningly (and cheaply) deploys this normalising power, it follows that the 'global political economy ... is a disciplinary society'.[15]

The basic feature of the neo-Gramscian approach, as elaborated by Robert Cox and Stephen Gill, is its holism.[16] It draws on several holistic sources. Foremost of course is Antonio Gramsci himself, for whom the 'organic' features of social relations were primary. Other sources include Karl Polanyi, who insisted that the market must be understood in relation to societal totalities, and Fernand Braudel, who focused on all-encompassing historical structures.[17] It should come as no surprise, therefore, to find that Cox describes his position as that of 'a "holist" who considers individual events as intelligible only within the larger totality of contemporaneous thought and action'.[18] Similarly, Stephen Gill asserts that 'the global system needs to be conceived as a totality'.[19] The 'world order' or 'global political economy' in this context represents the whole.

In the absence of an explicit theoretical defence of holism, it is necessary to reconstruct the principal holist themes. One theme of holism is a favourable stance toward intellectual comprehensiveness. Cox commends historiography that is 'comprehensive both in the way [it] conceives the world as a structural totality and in [its] manner of linking together the different aspects of social life'.[20] Gill promises 'a broad-based and more integrated perspective' in place of 'a narrow political economy approach'.[21] Another theme of holism is its enmity to 'atomism'. Cox regards 'problem-solving' theory as founded on an illegitimate carving-up of the fundamentally interconnected global order.[22] Gill enjoins us to accept that 'the basic unit of analysis' is not any kind of individual but 'the ensemble of social relations configured by a social structure'.[23] A third theme is that one of the main explanatory devices used by holists is the proposition that a change in one area of the system causes change in all others. Gill endorses this when he identifies 'a key issue' as the idea that 'a change in thinking is a change in the social totality and thus has an impact on other social processes'.[24] Indeed, this axiom of necessary interconnection reappears as the basic methodological guide.

Perhaps the most important feature of holism, and the most unconvincing, is its *a priori* character, which is to say that holists assume the quality of wholeness must be posited in advance of theoretical explanations or empirical investigations. This accounts for the way in which it tends to appear as an initial assertion rather than as a final conclusion. Gill has frequent recourse to this assumption: 'First, and most fundamental', he asserts, theorists 'must analyse it [the world order] as a whole', and he continues, 'our ontology must be founded upon the idea of global social formation'. Later he states that the global political economy is 'the fundamental concept of our ontology'.[25] To invoke fundamentals in this way is merely to reassert the ultimate priority of the whole.

Starting from a holist position, the conceptual apparatus of the neo-Gramscian approach may be easily explicated. Incorporative principles such as 'hegemony', 'intersubjectivity', or 'organic' are conceptually deployed to secure the cohesion of totalities designated as 'structure', 'ontology', or 'historic bloc'. Within the global order there may be differing 'dimensions' or 'levels' consisting of political, cultural and economic spheres, but these are not separate; rather they are dialectical moments that are related internally rather than externally as autonomous institutions. Against the institutionalists, the claim is that the spread of markets and the distribution of political institutions are encompassed in an order that is shaped by the 'social forces' that arise from circuits of capitalist production and ideology. The key dynamics that integrate the world into a whole thus originate in the spheres of global production and universalising ideology.

Central to accounting for the cohesion of the totality is the theory of hegemonic incorporation. At the most basic level it posits an intersubjectivity or a certain set of meanings and beliefs that are widely shared in consensual fashion, at least among elites, throughout the world. At an intermediate level the theory posits an ideology, understood as the set of ideas that serve to prevent subordinate groups from creating an alternative to the global order. Ideology is thus conceived as a global cement. Finally, in the upper levels, the theory posits a social bloc, based on the class and other beneficiaries of global incorporation, itself integrated in a similar fashion. Such incorporation serves to integrate the various dimensions of the global order, forming a mutually reinforcing political, economic and cultural structure.[26]

What happens if surveillance is added to the whole? In the following example, it is understood as a further means of securing hegemonic incorporation, a power that prevents rather than facilitates differentiation. The authors of a work on 'cybernetic capitalism' take this stance. In the latest (cybernetic) stage of capitalism, with the convergence of computer and telecommunications technology in the context of a crisis of Fordism, information becomes the unifying filament of the totality. Surveillance becomes a means for achieving the ultimate domination of capital: 'Through the "information revolution" capital invades the very cracks and pores of social life ... the reach of capital is extended throughout society' with the result that 'society as a whole comes to function as a giant panoptic mechanism'.[27]

The lesson to be drawn is not that the gaps in even the most heroic of attempts at holistic theorising need filling. It is that there is good reason to avoid the tendency to incorporate everything in a unity and systematically to neglect differentiation.[28] Surveillance is a source of the plurality of the global order and it encourages a certain reflexive contingency. A differing approach would seek to escape the poor choice of either a complete neglect of surveillance power or an overwhelming global discipline. It would aim to conceive the global political economy in terms other than the separation of institutions or their hegemonic amalgamation.

CONSTITUENTS OF A GLOBAL ORDER: DISCIPLINE, ORGANISATION AND REFLEXIVITY

In this section I present some suggestions for reconceiving the global political economy to make room for the possibility that surveillance plays a constitutive role. As a point of departure, consider one of the most vivid documentary accounts of the place of surveillance in modernity. It is the travelogue of the first Chinese envoy to the modern West.[29] Kuo Sung T'ao journeyed to London in 1876–7 and recorded his observations of the European 'Way'. What he wrote is of particular interest because he found himself in the singular position of viewing modernity as an outsider. The influence of western culture at that time on Chinese officials was still negligible. Kuo was also a well-educated man in the Chinese classical tradition and a privileged ambassador from a self-confident civilisation. As an outsider, Kuo was able to perceive what to insiders is the overly-familiar 'furniture' of their daily lives. As a well-educated diplomat, he could record and reflect on the experience. As a privileged visitor, he was able to inspect at first hand powerful institutions.

The West, Kuo argues, has a Way in which 'people have been competing with each other with knowledge and power', that has endured, expanded and that 'assists them in the acquisition of wealth and power'. What is this Way? The majority of Kuo's observations are of governmental institutions, which he characterises as 'well-ordered, enlightened, and methodical'. Of a school, he records the 'completeness' of its planning and administration, while of a prison, he notes the 'undeviating impartiality' of its discipline. A particular fascination of his seems to have been for zoological and botanical gardens, where he enthuses over the 'exotic flowers, curious plants, rare birds, and marvellous animals' collected together from around the globe. Kuo also discusses western economic practices and, as with government, he emphasises that their commercial 'methods are exact', their regulations are 'orderly and dignified', and this provides a 'firm foundation' for their wealth and power. He describes many practices that stimulate commerce, such as the provision of meteorological offices, the way that individuals 'scheme with all their might for the profit of their country', and the use of timetables: 'so accurately are all the multifarious details arranged that there is not the least fear of mistake'. Kuo recommends to his fellow officials that since their Way makes Europeans so wealthy and strong 'we must study ways of dealing with them'.[30]

What Kuo has documented are some of the chief features of surveillance. Techniques of surveillance that he describes include the timetables, with their accurate classifications and standardised time, the meteorological offices that gather regular observations from diverse sites into one centre, and even zoos and botanical gardens that collect and consolidate specimens from a diversity of places in one site. Of the creations of surveillance, there are three aspects. First, Kuo was much impressed by the level and pervasiveness of discipline: the impartial, complete, orderly, methodical, exact and accurate conduct of individuals in the modern West. Second, Kuo tended to single out the high level of organisation in the West, as evidenced in such matters as planning, administration and commerce. Finally, if we switch attention to the context of the dispatch itself, Kuo's memorandum itself is an example of a distinctly modern reflexivity. In that period

Chinese officials were beginning to mobilise information in order to monitor the West, beginning with the acquisition of western atlases and the sending of emissaries. Together, discipline, organisation and reflexivity contribute to a dynamic and distinctive Way that has become our modern global society.

Social discipline, as Michel Foucault has shown, is a product of surveillance concentrated within institutions of incarceration. Kuo also noted the importance of methodical conduct in the modern West and explicitly described the schools and prisons that he visited as highly disciplined. By analogy, global discipline has been seen as a product of global surveillance. Foucault selected the image of the Panopticon as the image of surveillance in action. The plan for a Panopticon prison had originally been formulated by Jeremy Bentham, whose aim was to design a means for the inexpensive, efficient and reformative management of criminals. As he wrote in the *Outline for the Construction of a Panopticon Penitentiary House*, it would consist of an enclosed space in which inmates are separated into individual cells arrayed around a central watchtower. By this architecture a single and central source of supervision would be able to observe each of the inmates all of the time. Bentham believed that supervision of this sort would instil in those subject to it 'the sentiment of an invisible omnipresence'.[31]

Foucault redescribed that presence as a 'totalising gaze' with its source in surveillance. Briefly, in a panoptic system, the objects of observation are individual bodies, their conduct made transparent by distributing, partitioning and positioning in such a way as to secure their individualisation. The observer, meanwhile, is able to monopolise the means of monitoring by its centrality and opaqueness in relation to the peripheral transparency of the observed. A panoptic system thus enables continuous and unidirectional supervision turning the individual into 'the object of information never a subject in communication'. The power of surveillance operates as 'a subtle calculating technology of subjection' on individuals by means of an internalised and penetrative power, discipline, which imposes on its objects a methodical docility.[32]

When a leading poststructuralist writes of 'imposing international purpose' by 'disciplining global conduct', portraying it as normalising and silencing but produced by a 'strategic reserve' of regions of undecidability and ambiguity,[33] it would seem that disciplinary power is being invoked simply as a distasteful principle of global life, unconnected with surveillance and incarceration. A complementary stance would identify surveillance solely with discipline, global surveillance with global discipline, the global order with a prison. One contribution asks 'how it was possible to construct through disciplinary power the underdeveloped state'? The unsurprising answer depends on analogy: 'much like the prison would rehabilitate the criminal to become a useful and productive citizen, agencies of political development – the IMF, the World Bank, the UN, the Ford Foundation, and university systems – would seek to reform the undeveloped state, the traditional or backward society, in order to change it into a productive, modern nation-state'.[34] The analogy supposes that because the incarceration of individuals and the stimulation of societies each depend on modes of surveillance, both are analogues of discipline. To take a third example, it may be admitted that 'obviously, in an anarchical society there is no central watchtower to normalise relations, no panopticon to define and anticipate delinquency' and yet proposed that espionage by means of technical intelligence systems, the panoply of electronic communication monitoring, constitutes a 'normalising, disciplinary, technostrategic power of surveillance'.[35] Again there seems to be the residue of an incarceral analogy.

A number of general features of panopticism suggests its inapplicability as a paradigm of surveillance in the global order. Most obviously, it is doubtful that a sentiment of an invisible omnipresence is universally held. Foucault's description also makes it plain that the targets of disciplinary power are individual bodies and not, for example, Third World states. It also indicates a monopoly of monitoring, whereas in the global order multilateral monitoring is more prevalent. Finally, in panopticism discipline is an imposition of docile conduct rather than the empowerment associated with surveillance in other arenas.

As an alternative it is necessary to start by recognising that global surveillance is too decentralised to effect a global discipline. Instead, discipline should be thought of as focused in highly localised sites, joined in networks that may stretch across the globe. For example, expert systems, themselves partly constitutive of surveillance, are also shaped as networks of professionalised agents through individualised discipline. Expertise is itself concentrated information. In some cases this enables the creation of long-distance networks of permanently stationed agents. Such arrays are part of what gives modern society its globalising tendency. By contrast, in pre-modern society only the Church achieved this on an extensive scale, for provincial officials of empires were virtually autonomous. From the time of Renaissance, states and merchants' organisations began the practice of stationing agents abroad in order regularly to report back information and act in contingencies. This in turn required a supply of reliable and knowledgeable agents who would be able to maintain their integrity, loyalty and efficiency at great distance from their controller. These are the origins of a modern standing diplomacy, which emerged as a system of monitoring through regular dispatches. In this case, the site of discipline and surveillance is the professional diplomat's career. Self-discipline was a quality stressed in one of the most widely used how-to manuals, *The Art of Negotiating with Sovereign Princes* (published in 1716). The aim of this advice book was to inform the reader of the disciplines, skills and expertise required to be a successful negotiator. It opens by identifying the problem to be rectified, complaining that good negotiators 'are more rare with us because we have no discipline, or certain rules, established for training fit persons in the knowledge of such things as are necessary'.[36]

The organised quality of modern life, also emphasised by Kuo, is a second aspect of the facilitative power of surveillance. The global political economy is a highly organised entity in that it is composed in part of distinct organisations such as states and corporations, and less formal organisations such as international regimes. But it is also organised in a broader way. Its political institutions interpenetrate to the degree that it may be possible to conceive a decentralised, turbulent, but nevertheless organised global polity.[37] A focus on surveillance directs attention to the source of organisation not in the formal structure of organisations (formally of course there is no 'global polity' only an anarchy) but to their infrastructure. It is not conventional among students of the global political economy to regard the global polity as having an infrastructure in surveillance power. Conversely, the historical sociologists that have brought the question of collective power into discussions of the state have tended not to discuss the global polity in similar terms. There is a need to bring the two together in order to explore the role of surveillance in constituting the global polity.

Michael Mann refers to collective versus distributive dimensions of state power. The former consists of 'the capacity of the state actually to penetrate civil society, and to implement logistically political decisions throughout the realm'. The state is nothing in itself other than a centralisation of the capacity to assess and tax income, to monitor and enforce rules, to influence economic activities, and to recruit and directly employ a staff. All these practices either first appear or grow rapidly in modern times with the rise of the nation-state. In Mann's 'organisational materialist' approach, these powers depend on 'logistical techniques', of which he mentions such constituents of surveillance as central co-ordination of specialisation, literacy (which enables transmission and storage of data), coinage, statistics, and accountancy.[38] Mann is interested in the degree to which such organisational powers become centralised and thus constitute state power. A complementary focus would attend to the degree to which the practices spread across the globe, create capacities of mutual interpenetration, and constitute new domains that form the global polity.

Similarly, Charles Tilly discusses the distinctively modern practice of 'direct rule'. All traditional states ruled indirectly, and intermittently, through local notables and cannot be said to have engaged in 'governing' as that is now understood. With the emergence of direct rule, which Tilly dates from the eighteenth century, modern states gained 'access to citizens and the resources they controlled through household taxation, manpower conscription, censuses, police systems and many other invasions of small-scale social life'.[39] All the examples that Tilly gives here rely on accumulated surveillance. In the global polity, the ability to gain direct access to distant domains is a part of the stretching of governing practices associated with the globalisation of politics.

The most comprehensive theorisation of the sources of modern organisation in surveillance is that of Anthony Giddens. Defining surveillance as the collation of coded information and the direct supervision of activities, Giddens traces how the dependence on forms of tallying of objects and persons by traditional states 'took-off' in modern states with a massive expansion of surveillance resources. State documentation, for example, started to grow in the sixteenth century after the influence of print culture enabled the codification of law, the reproduction of records and reports and, particularly from the nineteenth century, the burgeoning of statistics. Like other modern organisations, states depend on regularised information which they use in supervisory settings to displace traditional practices, to co-ordinate activities and to administer extended areas. The practice of administration produces state organisation from surveillance resources. Giddens discusses several effects of intensified administrative power: monitored borders, on which are concentrated an apparatus of customs, guards, and passport co-ordination; 'internal pacification', based on the monopolisation of organised violence by the state and the rise of military and police forces; and 'system integration' of extended areas into a unified territory. A bounded, pacified and integrated domain defines the modern state. Again, administrative practices also extend among states to produce not a distinct territoriality but a series of overlapping domains of multilateral monitoring that define the global polity.[40]

The rise of surveillance structures in global modernity enables an intensification of reflexive action.[41] Kuo's reporting on the West was one small example of conduct constituted by systematic monitoring. More generally, to speak of global movements 'acting', or of the 'acts' of large scale organisations such as multinational corporations and nation-states, is to refer to the way that such bodies continuously monitor the conditions and outcomes of their conduct. As a consequence, there is a global but decentralised network of multilateral monitoring formed by the assiduous evaluation of strategies, the mobilisation of precedents, the composition of plans, the incessant watching of problems. What this points to is not, as in many theories of action in the global political economy, a matter of obedience to norms and rules, nor of determination by ideology. Instead it points to the crucial role in facilitating action of an incessant flow of concentrated information. Surveillance power, in this aspect, has significance as both a condition and an outcome of intensely reflexive action.

The view that modern modes of acting are distinctively reflexive derives in part from Max Weber's account of the way that social practices have become subject to heightened 'rationality' in modern society. Since Weber's ideas are open to differing interpretations, it is worth pointing out that some versions would leave little space for surveillance. Those who would treat rational action as a purely 'ideal' or mental process of course detach their theories from the material logistics of surveillance. Those who regard rational action as a monolith that incorporates within itself all areas of life neglect the diversity of sites in which surveillance operates and its variety of forms. Those who view rational action as formal and essentially regulated supply no account of a dynamic force behind the process, such as might be entailed in the mobilisation of surveillance. By contrast, it seems more promising to pursue a version that recognises a certain plurality, concreteness, and dynamism to rational action, such as may be sought in the relation of reflexivity to surveillance resources.

Take the case of instrumental action. This is the type of rational action studied in theories of markets and policy, for instance, by the self-labelled 'rationalist' approach to the global political economy.[42] It is 'instrumental' in the sense that its mode of operation is the strategic pursuit of private preferences. A focus on reflexivity can illuminate some of its sources. Since it involves the practice of reckoning and ranking of preferences and available means, instrumentality depends on the widespread availability of counting and classifying devices, such as money or account books. Weber noted that a condition of instrumentally rational action is 'the degree in which the provision for needs ... is capable of being expressed in numerical, calculable terms'. Elsewhere he observed that in modern rational capitalism 'everything is done in terms of balances'.[43] The capacity to create numerical, calculable expressions that enable systematic comparison is precisely the power of surveillance. With this as a focus, the element of historical development enters – a phenomenon that tends to be bracketed in studies of instrumentally rational action.

The case of regularised action is somewhat similar. This is the kind of action whose mode is the following of rules and laws.[44] The key here is not instrumentality but consistency. A trade regime, for example, provides a framework for mutual and reciprocal recognition of rights and duties, where mutuality or reciprocity imply consistent action. It is unlikely that rational consistency on a widespread scale could subsist without the codification of rules and law, along with their interpretation by expert systems of legal professionals, that provides a memory of precedents inscribed in accessible records. Cascades of files or cases are crucial resources of enhanced information. Accumulated surveillance thus enables not only the continuous monitoring of the adequacy of a regime and deviations from it, but it also facilitates reflexive rule-following itself.

The reflexivity of governing is particularly important in shaping the global political economy. The problem of governance in globalising society is central to the study of the global political economy. Robert Gilpin treats it as the issue of how a dominant state 'emerges to give governance to the international system'; Robert Keohane and Joseph Nye's formula concerns the role of institutionalised 'governing arrangements' in managing an increasingly interdependent world market; Hedley Bull held a somewhat wider focus on the creation of multilateral 'order' in expanding international society; finally, Stephen Gill and David Law single out the hegemonic rule of an 'emerging transnational historic bloc', and its capitalist class, 'in an age of internationalisation of production and exchange'. These examples indicate the range of established current positions on the question of governance in globalising society.[45] For institutionalists, governing practices are understood as secondary to the institutional location of authority in states or international institutions. For holists, governing practices are secondary to the sources of hegemonic incorporation. By contrast, the term 'reflexive governing' is intended to identify the actual practices of modern information-dense governance.

Reflexive governing is a modern enterprise that monitors problems, penetrates domains and formulates programmes as information is accumulated. It also alters the character of governance in the light of new information.[46] Reflexivity may include 'policy', whether instrumental or consistent, but is not reducible to overt policy. It is decentralised but coheres into the basis of a global polity by virtue of its multilateral interconnections. It provides the conditions for the co-ordination of the separate institutions of the global political economy. An example of reflexive governance is the mutual monitoring of populations, the systematic and ongoing evaluation of comparative levels of welfare and security. It enters constitutively into 'internal' governance. When separate populations are visualised and informable, an 'international' domain emerges in which problems and programmes are reflexively compared. Extended 'competitiveness' is shaped by the intensified reflexivity of modern governance. This kind of governance is not a repressive or a constraining framework but rather a dynamic force that helps to construct modern globalising society.

Surveillance power enables reflexive governance. The capacity to mobilise information from great areas and to cascade it encourages an ever-greater degree of reflexive monitoring of interconnections across diverse locations. The spread of informable practices into a global society or global economy creates a wider area for problems or turbulence that require reflexive management or solution. All these capacities of surveillance depend on expertise of a reflexive or self-monitoring kind, which means that the spread of this modern form of governance would not be possible without the enhanced capabilities of the governed themselves. In high modernity, reflexive governance is becoming intensified – its procedures and responses are expanding and its operation is extending across the globe. This acceleration of networks of multilateral monitoring furthers the role of surveillance in globalising society.

CONCLUSION

In studies of the global political economy, surveillance tends to be either inflated or marginalised. It is either everywhere or nowhere, either an encompassing global discipline or a mere epiphenomenon. This chapter has outlined an approach that avoids these mirror-image misunderstandings. It requires a reconceptualisation of both surveillance and the global political economy. Surveillance is a collective enabling power whose origins are the accumulation of information. What does it enable? Not a 'global discipline' as in the model of a universal prison, but decentralised long-distance networks of professional experts. It also enables the mobilisation of organisational resources thereby shaping the organised character of the global political economy. Finally, it enables the ongoing development of networks of multilateral monitoring, and through reflexive governance, the related formation of the global polity.

Reconceiving the global political economy means asking how it is constituted. One answer is that the relations between global market and state institutions compose the global political economy. Another gives emphasis to the incorporating forces of a hegemony rooted in global production and universal ideology. But the global political economy is also fundamentally constituted by the practices entailed in reflexive governance, organisational extension and disciplined expert systems.

Notes

1. The 'three faces' are from Steven Lukes, *Power: A Radical View* (London: Macmillan, 1974). Cf. Stephen Gill and David Law, 'Global Hegemony and the Structural Power of Capital', *International Studies Quarterly*, Vol.33 (1989) pp.475–99.
2. From John Brewer, *The Sinews of Power: War, Money and the English State 1688–1783* (London: Unwin Hyman, 1989).
3. Edward Rolf Tufte, *Envisioning Information* (Cheshire, Conn.: Graphics Press, 1990).
4. My understanding of 'bias' is indebted to Edward Comor, 'Harold Innis and Marxist Methodology' (unpublished).
5. Elizabeth L. Eisenstein, *The Printing Revolution in Early Modern Europe* (Cambridge: Cambridge University Press, 1983).
6. Bruno Latour, 'Visualisation and Cognition' in H. Kuklick and E. Long (eds), *Knowledge and Society*, Vol.6 (Greenwich, Conn.: JAI Press, 1986).
7. Eisenstein, *Printing Revolution*.
8. See Anthony Giddens, *The Constitution of Society* (Cambridge: Polity Press, 1984) pp.145–51.
9. See Robert Keohane, 'International Institutions: Two Approaches', in his *International Institutions and State Power* (Boulder: Westview, 1989).
10. Joseph Addison, 'The Royal Exchange' in Richard Steele and Joseph Addison, *Selections from the Tatler and the Spectator* (Angus Ross ed.) (Harmondsworth: Penguin, 1982) pp.437–40.
11. J.G.A. Pocock, *Virtue, Commerce, and History* (Cambridge: Cambridge University Press, 1985).
12. Cf. Martin Wight, *International Theory* (Leicester: Leicester University Press, 1992).
13. Robert Keohane, 'The Demand for International Regimes' in his *International Institutions and State Power*, p.120. On the role of monitoring in international organisations, see Ernst B. Haas, *When Knowledge is Power: Three Models of Change in International Organisations* (Berkeley: University of California Press, 1990).
14. See most recently Christopher Chase-Dunn, *Global Formation* (Oxford: Blackwell, 1989).
15. Bruce Andrews, 'The Political Economy of World Capitalism: Theory and Practice', *International Organisation*, Vol.36 (1982) pp.135–63 at p.136, p.156 and p.157 (emphasis in original).
16. See most recently Stephen Gill (ed.), *Gramsci, Historical Materialism and International Relations* (Cambridge: Cambridge University Press, 1993).
17. See Fred Block and Margaret Somers, 'Beyond the Economistic Fallacy: The Holistic Social Science of Karl Polanyi' in Theda Skocpol (ed.), *Vision and Method in Historical Sociology* (Cambridge: Cambridge University Press, 1984).
18. Robert Cox, 'On Thinking About Future World Order' in *World Politics*, Vol.38 (1976) pp.175–96 at p.182.
19. Stephen Gill, 'Historical Materialism, Gramsci, and International Political Economy' in Craig N. Murphy and Roger Tooze (eds), *The New International Political Economy* (Boulder: Lynne Rienner, 1991) p.70.

20. Cox, 'Future World Order', p.176.
21. Gill, 'Historical Materialism', p.57.
22. Robert W. Cox, 'Social Forces, States and World Orders' in Robert O. Keohane (ed.), *Neorealism and Its Critics* (New York: Columbia University Press, 1986).
23. Gill, 'Historical Materialism', p.56.
24. Ibid., p.59.
25 Ibid., p.61 and p.73.
26. See Robert W. Cox, *Production, Power, and World Order* (New York: Columbia University Press, 1987).
27. Kevin Robins and Frank Webster, 'Cybernetic Capitalism: Information, Technology, Everyday Life' in V. Mosco and J. Wasko (eds), *The Political Economy of Information* (Madison: University of Wisconsin Press, 1988) p.54 and p.72.
28. On holism see Geoffrey Hawthorn, *Plausible Worlds* (Cambridge: Cambridge University Press, 1991).
29. Kuo Sung T'ao et al., *The First Chinese Embassy to the West* (J.D. Frodsham ed. and trans.) (Oxford: Clarendon Press, 1974).
30. Ibid., pp.72–3, p.43, p.8, p.13, pp.35–6 and p.43.
31. Jeremy Bentham, *Outline for the Construction of a Panopticon Penitentiary House* in Mary Peter Mack (ed.), *A Bentham Reader* (New York: Pegasus, 1969) p.194.
32. Michel Foucault, *Discipline and Punish: the Birth of the Prison* (New York: Vintage, 1979) p.200 and p.221.
33. Richard K. Ashley, 'Imposing International Purpose: Notes on a Problematic of Governance' in E.-O. Czempiel and J.N. Rosenau (eds), *Global Changes and Theoretical Challenges* (Lexington: Lexington Books, 1989) p.226.
34. Deborah S. Johnston, 'Constructing the Periphery in Modern Global Politics' in Murphy and Tooze, *New International Political Economy*, p.151 and pp.165–66.
35. James Der Derian, 'The (S)pace of International Relations: Simulation, Surveillance, and Speed' in *International Studies Quarterly*, Vol.34 (1990) pp.295–310 and p.304.
36. Francois de Callieres, *The Art of Diplomacy* (H.M.A. Keens-Soper and K.W. Schweizer eds) (Leicester: Leicester University Press, 1983) p.66.
37. Cf. Cox, *Production, Power, and World Order* on 'the internationalisation of the state'.
38. Michael Mann, 'The Autonomous Power of the State: its Origins Mechanisms and Results' in his *States, War and Capitalism* (Oxford: Blackwell, 1988).
39. Charles Tilly, *Coercion, Capital, and European States* (Oxford: Blackwell, 1992) p.25.
40. Anthony Giddens, *The Nation-State and Violence* (Cambridge: Polity Press, 1985) p.86.
41. See Anthony Giddens, *Consequences of Modernity* (Stanford: Stanford University Press, 1990).
42. Keohane, 'Two Approaches'.
43. Max Weber, *Economy and Society* (Berkeley: University of California Press, 1968) p.85; *The Protestant Ethic and the Spirit of Capitalism* (London: Unwin, 1976) p.18.

44. Friedrich Kratochwil, *Rules, Norms, and Decisions* (Cambridge: Cambridge University Press, 1989).
45. Robert Gilpin, *War and Change in World Politics* (Cambridge: Cambridge University Press, 1981) p.198; Robert Keohane and Joseph Nye, *Power and Interdependence* (Boston: Little, Brown, 1977) p.5; Hedley Bull, *The Anarchical Society* (London: Macmillan, 1977); Stephen Gill and David Law, *The Global Political Economy* (Brighton: Harvester, 1988) p.65.
46. But cf. Michel Foucault, 'Governmentality' in G. Burchell et al. (eds), *The Foucault Effect: Studies in Governmentality* (Chicago: University of Chicago Press, 1991); Peter Miller and Nikolas Rose, 'Governing Economic Life' in *Economy and Society*, Vol.19 (1990) pp.1–31.

Part II
Case Studies

5 Communication Technology and International Capitalism: The Case of DBS and US Foreign Policy[1]

Edward A. Comor

Over the past two decades, United States officials have sought an increasingly liberalised international communications environment. New technologies – directly involving computers and telecommunications – are now increasing the capacity of corporations and governments to control and profit from the production and distribution of 'non-material' information and entertainment products and services. Because of the high costs of producing the initial product and the relatively minimal costs of its reproduction, the production and distribution of these non-material commodities is increasingly becoming the domain of those possessing the capacity to control the product's or service's distribution after the initial point of production. Whether it is a new financial service, a computer database or a pay-per-view movie channel, new communication technologies have enabled corporations to substantiate the enormous overhead costs required to develop these products and services and to bring them to market at an unprecedented pace.[2]

Some of the leading proponents of these developments have been US state officials. While US world leadership in the promotion of these activities has been apparent since 1945, trade issues concerning them have come to a head over just the past decade.[3] The reason for this intensification is due in large part to the relative decline of the domestic US economy. The global institutional–legal reforms promoted by both US officials and transnational corporations – involving both multinational (i.e. the GATT) and bi- and tri-national trade agreements (i.e. the Canada–United States FTA and North American FTA) – have thus been matters of some urgency for those recognising America's economic future to be intimately linked with the uninterrupted growth of its strongest economic area relative to Japan and the European Community – its information and entertainment industries.[4]

This chapter will evaluate these efforts and their implications by focusing on one of the most important new international communication technologies – the Direct Broadcast Satellite (DBS). In brief, a DBS 'system' is made up of a ground station that processes and radiates a signal to a satellite in geostationary orbit 36 000 km above the earth's surface. There, the signal is reprocessed and amplified for the downlink stage. Finally, the system involves a ground receiver (the 'dish' or 'flat-plate') that collects the signal and processes it for viewing on a television screen. DBS systems are unique because of the tremendous power used to amplify signals for the downlink. In general, the more power used, the smaller the size of the ground receiver required to receive signals within the satellite's coverage area (often referred to as the satellite's 'footprint'). These receivers can conceivably be as small as one-foot in diameter and can be attached to a window sill.

In order to evaluate the relationship of DBS developments to US foreign policy, I will proceed as follows. First, in the section called 'Communication Technologies: Rolc and Context', I will attempt to define US foreign policy in the field of communications. This section will also make a number of theoretical points in an attempt to contextualise much of what follows, including some speculation as to the role played by communication technologies in relation to international capitalism. Then, in 'The Status of DBS', a brief overview of ongoing international DBS developments will be presented. Following that, 'Barriers to International DBS Video Services' will review some of the most significant hurdles now facing DBS plans and related US foreign policy interests. Then, under the heading 'DBS and US Foreign Policy', I will focus on the explicit and implicit relationships linking American policy and international DBS developments. Finally, in 'Conclusions', I will relate some of the theoretical points made earlier, with the implications of DBS and related developments for both United States' interests in particular and international capitalism in general.

COMMUNICATION TECHNOLOGIES: ROLE AND CONTEXT

From the outset, it is important to contextualise the relationship between communication technology in general and US foreign policy in particular. The first point to be made is that the United States government does *not* have a coherent, well-defined international policy in the field of communications. No one agency or department is responsible. Instead, depending on the specific issue, its context and the timing of its formulation, a range of political and bureaucratic actors – including Congressional committees, the Executive, the Department of Commerce, the Federal Communications Commission, the United States Information Agency, the State Department and the United States Trade Representative – may be involved to various and largely unspecified degrees.

Despite the enormity of the stakes involved in this policy area, this fragmentation of responsibility should not be of any great surprise. In part it is a reflection of the predominance of the day-to-day communication activities in the American private sector as communication and information producers and/or users and/or service providers. As a partial result of this predominance, the roles of various government agencies and departments can perhaps be best characterised as the *co-ordinators and mediators of private sector interests and conflicts*.[5]

Yet, despite the absence of *a* foreign policy in communications, a consistent *perspective* and *approach* to issues involving international communications has certainly been present. In sum, a common 'worldview' has consistently represented the perceived interests of 'a transnational, corporate business system and its ancillary functions'.[6] More concretely, this approach has involved three general activities:

1. Ongoing efforts to persuade all nation-states to become or remain direct participants in the international capitalist market system;
2. The protection or expansion of US public and private sector political and economic interests involving the unhindered or 'free flow' of information and entertainment; and,
3. The direct or indirect application of public and private sector communication resources in efforts to rollback the 'communist threat' (however defined).[7]

Underlying all of these activities has been the almost unchallengeable belief that,

1. Privately owned media operating in a competitive capitalist market are 'objective' and 'neutral' information managers;
2. A competitive free market in communications is the most efficient mechanism for meeting consumer/citizen demands, wants and needs; and,

3. The promotion of new communication technologies is the primary means through which the above ideals can and should be promoted.[8]

Perhaps the bottom-line underlying these activities and beliefs involves the relative strength of US communication interests such as AT&T, IBM and 'Hollywood' (the film and television industry) relative to international competitors, and a shared understanding that communication technologies will facilitate the development of new economic opportunities leading to greater national wealth. As Herbert Schiller has recorded, these beliefs have been and are being actively promoted by the United States government, often in ways quite contrary to the open, free-market and 'level-playing-field' principles used to substantiate them:

> These doctrines and practices ... have been supported by national diplomacy and sometimes force, as well as by an exercise of monopoly control of communication facilities wherever, and as far as, possible. In general, a very close co-operation between the government and the private corporate media in the pursuit of these positions has [characterised] ... the last four decades.[9]

On a more theoretical plane, one of the most remarkable features in the modern history of 'developed' capitalist countries has been the rapid decline of long-standing social institutions. Pre-capitalist mediators of social reality – such as the Church – have now largely been replaced by institutions such as science, democracy and the price system. Indeed, one of the key facilitators of this transition has been the emergence of the electronic media. Through its development as both the producer and supplier of domestic and international information and entertainment products and services, new ways of thinking about the social and physical world have been facilitated and/or encouraged. Technological developments involving these media have been perhaps most significant in that *they have themselves emerged as commercial institutions promoting their own utilisation*. Moreover, this utilisation has itself modified ways of thinking and acting.

As mentioned at the outset, a central dynamic here is the often-high economic and social costs involved in the initial development and application of new technologies. Someone, of course, must bear these costs and, at least initially, they will probably be met through the co-ordination of various interests. Having made substantial investments, these interests will tend to become actively involved in ensuring that the unused capacities of the technology in question are redressed. For similar reasons, the overhead costs of new technologies also tend to stimulate a general centralisation in control over management activities. If new technologies can accommodate this centralisation, transnational corporate activities are provided with an opportunity to become geographically diversified, resulting in at least the potential to reduce the risks involved in concentrating production and marketing activities in a relatively small number of locations. However, a contradictory result of this simultaneous spatial diversification *and* management centralisation is the emergence of a potentially more volatile international order. Because a war or economic crisis in one part of the world becomes more and more 'the business' of other parts of the world, local disruptions tend to produce magnified and immediate shocks to disparately located areas.[10]

In sum, the high overhead costs and the scale of capital integration involved in the development of new international communication technologies tend to stimulate these contradictory developments. And, as I will point out, the development and application of international DBS systems constitute a contemporary case study of these tendencies.

THE STATUS OF DBS

Compared to less powerful satellites, DBS systems possess important advantages. The first is the capacity for just one DBS transmission beam to cover one-third of the earth's surface. A second is that unlike terrestrial or cable broadcasts, DBS signals are largely accessible independent of the receiver's relative isolation or the extremes of surrounding terrain. A third involves its power output and its signals' subsequent reception direct to the home without the need for redistribution through costly and state controllable ground stations or cable connections. And fourth, the signal provides a significant qualitative advantage over terrestrial and most cable transmissions. While DBS's technical features now enable it to broadcast high definition television (HDTV) signals, most cable systems are still developing this capacity. From an economic perspective, one transmission station and one satellite can cover the geographic area that only hundreds of inter-linked ground stations and/or thousands of kilometres of cable wiring can match.

From the point of view of many transnational corporations, the primary advantage of DBS involves its potential to deliver a cost-effective international audience to advertisers. Moreover, because of the 'intimate' quality of the human voice and image entering one's home, DBS (as with all television systems) provides an effective means through which both a literate and non-literate population can be reached. It should also be mentioned that because of this 'intimate' characteristic an effective and immediate expression (and manipulation) of human emotion is a well-recognised political and commercial advantage relative to other forms of mass media. And, while international advertising and marketing executives now consider language differences to be *the* major barrier facing the development of global television systems,[11] DBS has the capacity for a receiver-controlled audio track to accompany a single visual broadcast. In addition to home television broadcasting, other commercial DBS applications include CD-quality sound deliverable to portable radios; digital television and radio for cars and trains; and innumerable international data services.

DBS-type systems also provide a competitive alternative for some forms of intra-corporate communications. DBS audio, data and visual services enable transnational corporations to avoid using publicly switched systems. Unaffected by terrestrial emergencies, crowding uncertainties, or price fluctuations, the intra-corporate market for private DBS systems will likely provide TNCs with unprecedented capacities to locate production and marketing activities anywhere in the world with relatively less concern regarding 'public' telecommunication infrastructures.

Because of its enormous coverage area most DBS signals inevitably spill over across national borders. Because of their non-terrestrial delivery, the 'jamming' of DBS signals is a difficult undertaking and, even if successful, is restricted to a limited reception area. The only other physical means of limiting DBS signal reception – short of destroying the satellite itself – is by restricting the distribution of reception equipment. However, because of their small size and low cost,[12] the ability of any state to restrict access to and use of DBS reception equipment is limited.

The 1990s is a formative decade for DBS. Since February 1989, the Luxembourg based Astra satellite has been beaming dozens of channels to most of Europe. Evidence of its success is its incorporation of compression technologies as a response to the growing waiting list of programmers seeking access to Astra's pan-European distribution network.[13] Over 20 per cent of the homes in Western Europe now receive satellite programming. By 1995, one study estimates that this will increase to almost 34 per cent and the number of European households receiving programmes directly through a reception dish will be over 11 million.[14] Perhaps most significantly, Astra's dominant position is driving Europe toward *de facto* industry transmission standards despite the industrial development plans of EC officials. In large part, this has been the result of Astra's utilisation of medium-power DBS technologies while most European regulations have only been applicable to high-power systems.

In Asia, the Japanese government, in partnership with the state-supported broadcasting company NHK, has spent, through to the end of

1990, approximately $2 billion (US) on the development and launching of DBS.[15] In 1991, 3.5 million people received these signals.[16] Also in 1991, the Hong Kong-based transnational corporation Hutchison Whampoa (HW) launched its DBS system – AsiaSat I – *designed* to defy international cross-border broadcasting conventions, potentially transmitting up to sixty channels into Asian countries. Its Star TV is the first service to provide 24-hour programming to the entire region. By providing advertisers with significant discounts, Star TV has secured approximately $300 million in advanced funds.[17]

In the United States, the apprehensions of many DBS license-holders regarding the economic viability of their plans – especially given the extensive cabling of urban centres – are largely over.[18] The reasons are both technological and regulatory. First, significant advances in the compression of signals has multiplied the number of channels that can be transmitted from a single satellite transponder. As a result, depending on the complexity of a particular signal, one satellite transponder can now broadcast up to eight channels. The second breakthrough involves advances in digital encryption, or 'scrambling' technology. This is important because it may reduce both the number of viewers receiving satellite signals through illegal descrambling equipment and the volume of video signals 'pirated' by entrepreneurs for redistribution.[19] And third, public and legal concerns over what have been called the anti-competitive practices of cable programmers have compelled the US Congress to re-regulate domestic cable companies. One result is that DBS operators now have access to America's most popular cable channels.

Thus, despite enormous overhead costs, the opposition of powerful cable interests, and the persistence of legal uncertainties, recent developments involving compression and encryption have probably made DBS systems economically feasible even in the North American market. But because of the traditional strength of terrestrial services in general, and cable in particular, it seems likely that the future success of DBS broadcasting will be in at least two areas.

The first relates to its capacity to deliver multilingual services across borders, irrespective of the regulatory status of receiving countries. These broadcasts will be supported largely by advertising and the direct corporate sponsorship of programming. The market for these mass transnational broadcasts will likely be areas that are relatively underserved by terrestrial systems. These include the 'peripheries' of both the developed and less developed world. In the former, the residents of isolated rural areas in North America, where large satellite dishes are already popular,[20] will likely be attracted to these services. In the latter, areas with an emerging 'middle-class' possessing the necessary income, time and interest constitute a probable market. Transnational corporations producing goods and services for a mass consumer market will probably support DBS systems targeted at these less developed countries.[21]

The other kind of DBS system that appears likely to succeed is a more complex, long-term proposition. Over the past decade, a proportional decline in television's share of corporate promotional spending has occurred.[22] The primary reason for this relative decline has been television's tarnished image as an ideal medium through which to reach targeted audiences. Even viewer monitoring technologies and related marketing strategies have been clouded by doubt as to their precision and cost-effectiveness.[23] From the perspective of the programmer, an increasing number of channels, the rising cost of programming and a general decline in advertising revenue has resulted in a potentially non-profitable broadcasting environment. It appears likely that advertisers will continue to demand 'more efficient buys' and that as a result, unprecedented pressures will mount on television programmers to guarantee the delivery of a specified group of targeted consumers to the advertiser-client. An almost immediate 'answer' may lie in DBS (and cable systems as conduit distributors) as the means of gathering a mass audience of consumers who share both a common interest and consumer lifestyle. This constitutes a logical *extension* of current cable-based narrowcasting practices toward what can be called 'transnational-narrowcasting'. Because of the capacity of DBS to scramble signals, a fee can be charged for each channel or programme received by the viewer. DBS may thus enable broadcasters to charge high advertising fees in return for its delivery of an 'interested' global audience (as the descrambling fee would likely indicate).

Established forms of narrowcasting may indicate great potential if narrowcasting is applied on a world scale. Examples such as an international pay-per-view equestrian horse channel, an international Japanese drama channel, an international bankers, channel, etcetera, could all attract a few thousand viewers from each country. But taken as an international whole, such an audience would represent a significant breakthrough for specialised sponsors. Moreover, the establishment of a specialised and interested international audience could well reduce the practice of 'zapping' advertisements, since the viewer of these narrowly-targeted programmes is presumably *also* interested in the sponsor's message (for example, viewers paying for 'German Dressage Newsmagazine' would presumably be interested in the sponsor's 'innovative' new horse medicine). Moreover, the direct participation of a pay-per-view audience would constitute a valuable tool in tracing the viewing and consumption patterns of an international clientele. Through this kind of knowledge, other direct forms of international promotion and marketing become possible.

However, questions regarding the American production and programming industry's capacity to fill up a two-hundred channel DBS system with the kind of programmes needed to attract viewers, advertisers and sponsors remain unanswered. Some television and satellite industry observers believe that an adequate supply of programming already exists for DBS to become profitable now that the US Congress has re-regulated the domestic cable industry, freeing up programming for the use of alternative distribution systems.[24] Others believe that appropriate new programmes will be produced once DBS capacities are established.[25] And still others consider that, at least initially, DBS systems will distribute existing but locally unobtainable channels from around the world to fill up unused transponder capacities.[26]

These predictions may well be overly optimistic for two reasons. First, it is questionable whether viewers who already have access to the still influential and expanding cable systems will choose to pay $700 (US) to receive DBS, especially if DBS programming does little more than offer what cable already provides.[27] Second, given the probability that DBS systems will have to offer the kind of niche and 'mass narrowcasting' programmes necessary to attract both specialised advertisers and audiences, questions must be asked as to the capacity of the production and programming infrastructure to both expand output and 'shift gear' into a *qualitatively new mode of production*. On this second point, timing (as goes the cliché) may well be 'everything'. Not only has the US Congress opened up the programming market to satellite broadcasters, the Federal Communications Commission has recently, for the first time, ruled that telephone companies can distribute television signals direct to homes through modified telephone lines. Thus, for DBS to become a core television service in North America, it appears that its 'window of opportunity' may be a decade at most. By the year 2004, most American and Canadian homes will likely be serviced by optical fibre television and/or telephone cables.[28] As noted above, it is DBS's capacity to accommodate new digital compression and encryption security technologies that has made it a viable service in the 1990s. However, if the production, programming and advertising infrastructure cannot meet the unique quantitative and qualitative needs of DBS before the end of the century (including perhaps the availability of some form of high definition television and reasonably priced digital television receivers), its potential to become a core video distribution technology may be undermined.

However, this cautious perspective may itself be flawed on two general counts. First, DBS systems are international in terms of both their technological capacity and their economic viability. The United States represents the central but nevertheless just one major television market. Much of the world, including areas that are attractive to 'elite' product advertisers such as parts of Europe and Asia, are relatively under-cabled or remain politically resistant to allowing the delivery of video services by traditional over-the-air broadcasters or telephone companies. Thus, despite DBS's questionable future as a core television system in North America, its potential in other parts of the world is promising. Second, DBS systems are now well beyond the planning stages in the US and elsewhere. As discussed, DBS can accommodate a variety of services other than the transmission of video entertainment and current affairs. The existence of unused transponder capacity therefore *may itself stimulate* the further development of intra-corporate and consumer market video and information activities, enabling TNCs and national corporations both to advance their private communication networks and extend their cultural influence over a disparately located workforce.[29] Moreover, services such as international home shopping networks can maximise the scale and efficiency capacities of the retailing industry.[30] In sum, a short-term shortage of *suitable* programming will likely itself stimulate the development of alternative international communication activities through DBS.

BARRIERS TO INTERNATIONAL DBS VIDEO SERVICES

As of 1992, the legal capacity to establish international subscription, pay-per-view and home shopping services is somewhat limited. However, *if* precedents such as the Canada–United States FTA and the North American FTA, EC television and telecommunications developments, and US-led efforts in the GATT on services and intellectual property continue *without the development of a significant and persistent international opposition*, the eventual establishment of an appropriate legal regime seems likely. In fact, some US producers are already selling the North American and world broadcasting rights to programming in recognition that television's future will be less defined by nation-state boundaries. As for the distribution of mass entertainment programming, despite (or perhaps as a result of) these advances, the European Community's resistance to US-made television and film products continues. This resistance is largely based on its defence of domestic markets in the face of significant US industry scale advantages in the production of information and entertainment products. In 1989, the EC adopted a Broadcasting Directive to ease the internal flow of television programming while also setting minimal content standards for advertising, sponsorship, right of reply, etc. Most offensive to both the United States government and American corporate interests was the introduction of European content quotas.[31]

Whether directly through the globalisation in demand for a particular service or product or, more generally, through the promotion of a consumerist lifestyle, DBS constitutes a significant advance in the aspirations of many transnational corporations. As Earl L. Jones Jr., Chairman and CEO of International Broadcast Systems Ltd, told a Congressional subcommittee,

> The privatisation of European broadcasting opens up new opportunities for American companies that advertise heavily on television. These companies are looking for global markets, and they need a global means through which to reach new customers on a cost effective basis ... [T]he more that American companies advertise abroad, the more of their products they will sell abroad, thereby contributing in a very positive way to the US balance of trade.[32]

Here it must be remembered that 'mass' audience DBS channels, at least initially, will be financed largely out of the pockets of transnational corporate advertisers. Because producers of mass consumer goods and services generally seek to communicate successfully with as large a demographically-defined audience as possible, DBS constitutes both an opportunity and a challenge. The opportunity is self-evident, but the challenge requires brief elaboration and it can be summarised in one question: how can a television programme and its advertising successfully resonate with a large and diverse cross-section of cultures?

On this we should remind ourselves of the history of Hollywood's predominance in international programming. That history involves an almost monopolistic control over distribution. This has been both accommodated by and has itself accommodated the development of an international mass demand for 'big budget' special-effects and/or fast-paced 'action' films. The predominance of these products, and their emulation through foreign 'copy-cat' productions, leads to a reaffirmation of the principle on which the domestic American entertainment industry itself developed. This principle can be simply stated: the less intellectually demanding and the more stimulating the product, the larger the potential audience. Of course the more competition that exists and the more focused the advertiser on a specific demographic market, the more likely it is that a film or programme will aspire towards a more 'defined' audience. Yet, given the multicultural and linguistic difficulties present, DBS will likely produce a bias towards more action- and special-effects-oriented programming despite their apparent unsuitability for small television screens. Moreover, with the mass potential of one-channel broadcasting to an international market (as opposed to 'just' the US market), the audience share required to attract adequate advertising dollars will be considerably lower for DBS. As a general result, the limited narrowcasting that has developed in the North American market as a result of increased competition for a relatively 'fixed' number of viewers may take some time to develop in larger, more 'flexible', transnational DBS markets.

Despite the technological and economic potentials of DBS, its emergence as the world's preeminent television delivery system, as indicated above, is being shaped by political, economic and cultural barriers. Because of the extremely high costs of establishing a DBS system, the vast majority of existing DBS operations have received extensive state support and/or funding.[33] From the satellite's construction through to the financial 'break even' point, the cost of establishing a commercial DBS system is at least $1 billion (US).[34] In addition, significant risks are involved in the implementation of DBS in areas where cable television already offers an unprecedented number of channels.[35] Finally, in anticipation that DBS's most lucrative pay-off lies in the uncertain development of HDTV, the absence of a common international HDTV transmission standard has deterred many large-scale investors.[36]

Nevertheless, DBS projects are well under way. Seen in their entirety, international DBS developments will contribute to the identification of an international 'middle-class' characterised by the predominance of 'lowest common denominator' programming. But also through DBS, an international elite will be located and defined in terms of particular consumer and professional identities. Because a programme will directly (and perhaps dearly) cost the viewer money, a qualitative expectation will be built into the broadcaster–viewer relationship. Because of this more specified and direct relationship, viewers will expect that changing demands will be quickly met. The critical 'consumer' will be encouraged and the exclusivity of viewers' shared interests and knowledge will stimulate a further cultural division of an international elite from larger political economic communities.[37]

DBS AND US FOREIGN POLICY

Given DBS's territorial reach, the successful advancement and application of DBS directly challenges *the* basic rule of international law – every nation enjoys the sovereign right to regulate its own domestic affairs, including its economic and cultural policies.[38] In sum, the goal of controlling both the production and distribution of international information and entertainment has brought successive US governments and a plethora of corporate transnationals together into a well-established united front despite the fragmented characteristics of US government communication policy. Thus, many of these transnationals have a direct or indirect interest in the success of DBS as a 'frontline' technology in ongoing efforts to liberalise international telecommunications and modify the conditions for international information and entertainment transactions.

As DBS becomes a central tool in the globalisation of marketing and consumption, two emerging social forces will be promoted: a mass of internationally disconnected 'consumers' perhaps defining themselves as 'middle class' and a more self-aware yet socially isolated international network of elite 'consumers' and 'professionals'. Both of these groups will perhaps stimulate the further internationalisation of the state while, because of DBS's one-way global communication tendencies, the capacity of workers (for example) to identify themselves and subsequently to organise internationally will likely remain weak relative to the technological and cultural integration capacities of elites. Moreover, the predominance of US cultural products will likely contribute to the ongoing leadership of US state officials in international institutions and negotiations involving the development of communication infrastructures, a 'free' trade in services, and a more developed intellectual property rights regime. We may also conclude that the commercial, short-term, acritical bias of the emerging transnational *mass* media products may contribute to the weakening capacity for sustained challenges to the free trade, free flow of information tendencies supported by many influential interests including the United States government. As a result, *the capacity to create the conditions needed to formulate a widespread and sustained counter-hegemonic vision by geographic and/or class peripheries will tend to remain underdeveloped.*

As discussed, the disparate and private sector-based character of US foreign policy in communications should not blinker us from recognising the presence of an ongoing policy consensus. To repeat, this has involved, first, ongoing efforts to persuade nation-states to become or remain direct participants in the international market system; second, the protection or extension of US political economic interests involving the 'free flow' of information and entertainment; and third, the application of communication resources in efforts to rollback the 'communist threat' (however defined). Of course this last point can now be modified in terms of threats to the so-called 'New World Order'. Nevertheless, international DBS systems can be seen to accommodate US foreign policy interests on at least four counts, *despite* the undefined nature of the relationship:

1. DBS represents a 'cutting-edge' technology in the establishment of international corporate freedoms to communicate whenever and whatever desired;
2. DBS represents a significant extension in unused communication capacity and subsequently unprecedented export opportunities for US entertainment and information producers, programmers and service providers;
3. As a result of both 1 and 2, DBS provides a significant advance in the capacity to promote both liberal democratic and consumerist values, both of which are central components in the future international political economic leadership of the United States; and
4. DBS systems, once established, may stimulate advancements in international agreements concerning trade in services and intellectual property rights, both of which have been long-standing priorities of US trade policy. In sum, once an international DBS system is established, reluctant nation-states either must live with the consequences or enter negotiations in order to pursue a more satisfactory relationship with external broadcasters.

CONCLUSIONS: DBS AND THE ROLE OF COMMUNICATION TECHNOLOGY

I have suggested that DBS will not only expand the marketing and advertising capacities of many transnational corporations but, more generally, the development of a globalised commercial television system will afford enhanced opportunities for influencing periphery cultures. By promoting a consumerist lifestyle, transnational advertising 'can severely weaken the effectiveness of formal government policies attempting to restrict transnational corporate access to national markets and to limit the commercialisation of their societies'.[39] Through DBS and related developments, a core–periphery model may be emerging in which the 'core' is best represented as transnational-capital and the 'peripheries' as a range of interests *including the nation-state*. Thus, on the one hand, the US government-US transnational corporate push for an international free flow regime constitutes an essential step towards America's economic recovery. On the other hand however, *the success of such a project could reduce US-based*

TNCs' dependency on the United States itself. Couched in the terms used by proponents of a New World Information and Communication Order, the eventual peripheralisation of the United States may paradoxically come about through the 'success' of US 'cultural imperialism'. As Jeremy Tunstall reminds us, the historic predominance of the US film and television industry is directly related to its absolute commercialisation. Thus, the international success of DBS will likely contribute to the elaboration of a world transnational corporate culture, rather than the 'US-Main Street' culture of the past.[40]

Perhaps the most important consideration in evaluating DBS involves its long-term influence on the intellectual (and consequently, organisational) capacities of viewers. Over time, we can assume that emerging identities will be modified as the core–periphery communication process is carried out through DBS but mediated by transnational sponsors. As the world becomes conceptually smaller, elites will tend to think themselves more familiar with the economic world around them. As a result, 'core' decision-makers may think themselves less dependent on the views of 'local' associates and decisions will tend to be made more immediately and more unilaterally. Consequently, crisis decisions will more likely involve the use of force. In Harold Innis' words 'improvements in communication have made understanding more difficult'.[41] With the emergence of international mass and elite DBS systems, a global consumerist identity perhaps will stimulate a false sense of understanding. Despite the public statements of US state and corporate officials who consider unimpeded market activities in information and entertainment to be core components of 'democracy',[42] this contradiction may itself be the most fundamental 'contribution' that global DBS will make to international relations.

Finally, what is equally remarkable is the contradiction emerging out of the United States' own foreign communications policy. With the globalisation of communication activities and the enormity of the costs and stakes involved in it, nation-states have become more and more the direct instruments of 'their' transnational corporations. As indicated above, both US domestic and foreign communications policy is structurally disparate and very much reflective of private sector conflicts and demands. This and the international competitive strength of US communication firms have been key stimuli in the leadership role of US state officials in the promotion of global reforms involving the 'free trade' of services and non-material commodities and directly related intellectual property agreements. In other words, US foreign policy in these areas appears to have become little more than the political expression of general transnational corporate interests and conflicts. Subsequently, political and military power has been used to protect and promote these interests. The result is that the United States and other national governments, in the words of William Melody, 'tie themselves more closely to the promotion of [the] corporate power of their TNCs ... [and as a result] they are ... reducing their own degrees of freedom to adopt domestic or international policies contrary to TNC interests'.[43]

In part as a result of the development of new international communication technologies – led by DBS – an international environment characterised by unprecedented interdependence and volatility is under construction. Without the development of adequate counter-weights to these tendencies involving spatial centralisation and temporal immediacy, the capacities of nation-states, corporations and people to meet the political, economic and environmental challenges of the future will continue to decline.

Notes

1. Thank you to Henry Comor, Leo Panitch, Ian Parker, Steve Patten, J. Magnus Ryner, Timothy Sinclair and Graham Todd for their helpful comments on previous drafts.
2. The service sector has grown to constitute approximately 60 per cent of advanced industrial countries' Gross National Products. From 1984 to 1990, the world market for telecommunications, computer products and services had an annual growth rate of 11 per cent, totalling $831 billion (US) in 1990. See Peter F. Cowhey, 'Telecommunications and Foreign Economic Policy', in P.R. Newberg (ed.), *New Directions in Telecommunications Policy*, Vol.2 (Durham: Duke University Press, 1989) pp.188-90.
3. One of the earliest and more explicit policy statements connecting international communications with US trade policy is a letter from then Secretary of State George Shultz to the Chairman of the Senate Committee on Foreign Relations (dated 21 September 1983). Copy available in Hearings on 'International Communication and Information Policy', US Congress, Senate Committee on Foreign Relations, Subcommittee on Arms Control, Oceans, International Operations and Environment. 98th Congress, 1st sess., 19 and 31 Oct. 1983, pp.4–5.
4. In an appearance before a US House of Representatives Subcommittee, the Motion Picture Association of America's Jack Valenti spoke on behalf of the International Intellectual Property Alliance. Representing seven trade associations involved in the production and distribution of movies, television, video and audio recordings, books and computer software, Valenti reminded the Subcommittee that these constitute more than 5 per cent of domestic GNP and, together, are the United State's largest trade surplus area (bringing in over $13 billion in 1988). 'Our world supremacy', warned Valenti, '... [continues to] evoke a kind of unruly response from too many countries'. See Hearings on 'Unfair Trade Practices', US Congress, House Committee on Energy and Commerce, Subcommittee on Telecommunications and Finance. 101st Congress, 1st sess., 1 March 1989, p.73.

5. Further evidence of the fragmented nature of the US policy-making structure can be found in persistent and long-standing suggestions of some domestic analysts and practitioners for a more cohesive policy-making process, including suggestions that a centralised Department of Communications be established. See for example Howard J. Symons, 'The Communications Policy Process' in P.R. Newberg (ed.), *New Directions in Tele-communications Policy*, Vol. 1 (Durham: Duke University Press, 1989) pp. 275–300. Personal interviews with public and private sector officials involved in US communications policy, held August–September 1992, confirm this structural characteristic.
6. Herbert I. Schiller, 'Is There a United States Information Policy?', in William Preston Jr., Edward S. Herman and H.I. Schiller, *Hope & Folly, The United States and UNESCO 1945–1985* (Minneapolis: University of Minnesota Press, 1989) p.287.
7. Ibid., pp.287–8.
8. For critiques of neoclassical economic assumptions underlying these, see Robert E. Babe, 'Information Industries and Economic Analysis' in Michael Gurevitch and Mark R. Levy (eds), *Mass Communication Review Yearbook*, Volume 5 (Beverly Hills: Sage, 1985) pp.535–46; and Ian Parker, 'Commodities as Sign-Systems' in Robert Babe (ed.), *Information and Communication in Economics* (Boston: Kluwer, 1993).
9. Herbert I. Schiller, 'Is There a United States Information Policy?', p.288.
10. Perhaps the most cogent example of this emerging international connectiveness, immediacy *and* volatility is the scale and magnitude of 'increasingly regular' international stock and money market 'panics'.
11. According to one survey of international advertising and marketing executives, transnational corporations are now seeking to centralise world promotional activities and, as a result, qualitative modifications in the marketing-advertising process are forecast. These anticipated changes will likely include a greater emphasis on the visual aspects of television advertising. See Donald G. Howard and John K. Ryans Jr., 'The Probable Effect of Satellite TV on Agency/Client Relationships' in *Journal of Advertising Research*, Vol.28, No.6 (December 1988/January 1989) pp.44–5. This point was confirmed in personal interviews with Les Margulis (International Media Director) and Eric Sheck (International Media Supervisor), BBDO Worldwide Inc., 21 April 1993, New York City.
12. In Japan, the best-selling television sets now have a built-in satellite tuner and an accompanying reception dish costing consumers an extra $300 (US). See Gale Eisenstodt, 'Peter Jennings in Japanese' in *Forbes*, Vol.147, No.3 (February 4, 1991) p.40.
13. Personal interview with Kenneth Donow, W.L Pritchard & Co., Inc., Washington DC, 4 September 1992. At least 20 million European homes now receive Astra and two million of these have purchased their own two-foot reception dish. See Jonathan B. Levine, 'This Satellite Company Runs Rings Around Rivals' in *Business Week (Industry/Technology Edition)* Issue 3199 (11 February 1991) p.74.
14. Ibid., p.75.
15. Gale Eisenstadt, 'Peter Jennings in Japanese', p.40.
16. Michael Westlake, 'Reach for the Stars' in *Far Eastern Economic Review*, Vol.151, No.22 (May 30, 1991) p.61.

17. Explaining these commitments, one advertising executive asks, 'Can I walk away from this, because it might be massive? ... If the price is right, the advertising buy is right'. Star's parent HW is targeting Hong Kong, Taiwan, South Korea, Singapore, Malaysia, Thailand, Indonesia, India, Pakistan and the Philippines with mostly US-produced programming, including an Asian version of MTV. See Michael Westlake, 'Reach for the Stars'. In mid-1993 Australian-based News Corporation International purchased a majority interest in Star TV. News Corp. also controls one of the most popular DBS services on the Astra system, British Sky Broadcasting. News Corp. also owns the DBS encryption technology most used in Europe, Asia and North America through its News Datacom subsidiary.
18. North America's first mass consumer DBS systems (Hughes Communications's DirecTV and Hubbard Broadcasting's USSB) will begin their transmissions in 1994. It has been thirteen years since the FCC first licensed nine domestic DBS companies to proceed with their business plans.
19. See, for example, Kenneth D. Ebanks, 'Pirates of the Caribbean Revisited' in *Law & Policy in International Business*, Vol.21, No.1 (1989) esp. pp.33–7.
20. As of 1992, there were approximately 3.7 million home satellite dish owners in the United States. See Satellite Broadcasting and Communications Association of America, 'Satellite TV Facts at a Glance' (SBCA pamphlet, 1992) p.1.
21. As mentioned above this kind of DBS service has already been established with Star TV.
22. See Florence Setzer and Jonathan Levy, *Broadcast Television in a Multichannel Marketplace* (Washington DC: FCC Office of Plans and Policy, 1991) pp.112–34. Also William Leiss, et al., *Social Communication and Advertising, Persons, Products & Images of Well-Being*, Second Edition (Scarborough: Nelson Canada, 1990) pp.189–91.
23. Backer Spielvogel Bates Media Department, *BSB Projections 2000, Media & Measurement Technology Predictions For The Coming Decade*, Internal Report (January 1991) pp.7–8.
24. Personal interview with Harry W. Thibedeau, Manager of Industry and Technical Affairs of the Satellite Broadcasting and Communications Association of America, Washington DC, 31 August 1992.
25. See 'Hartenstein Leads Hughes' DirecTV DBS Team' in *Satellite Communications*, Vol.16, No.2 (February 1992) p.17.
26. Personal interview with David Webster, President of the Trans Atlantic Dialogue on European Broadcasting, Washington DC, 11 September 1992.
27. $699 is the target price of a receiver and decoder unit for Hughes Communications and Hubbard Broadcasting high-power DBS systems DirecTV and United States Satellite Broadcasting (USSB), respectively. See Lloyd Covens, 'DBS: The View from Stan Hubbard' in *Satellite Communications*, Vol.16, No.2 (February 1992) p.14.
28. On US telephone and cable company activities involving optical fibre developments, see Richard A. Gershon, 'Telephone-Cable Cross-Ownership, A Study of Policy Alternatives' in *Telecommunications Policy*, Vol.16, No.2 (March 1992).
29. On preliminary developments in what is commonly referred to as 'business TV', see Mike Stevens, 'Getting Down to Business TV' in *Marketing*

(March 5, 1992) p.33 and p.35; and Geoffrey Shorter, 'Will Satellites Get You Closer to Customers?' in *Business Marketing Digest*, Vol.16, Issue 4 (1991) pp.57–64. In general terms, 'culture' here refers to the social environment within which people think and act. Moreover, culture is a dynamic concept representing an ongoing dialectic involving human activity and this social environment.

30. Personal interviews conducted by the author (see fn.5 above) indicate that extensive but confidential plans along these lines are now being made by one of the United States' largest retailers in partnership with one of its largest computer firms.
31. Jack Valenti and then US Trade Representative Carla Hills responded by lobbying the US House of Representatives to pass a resolution condemning the quotas as 'unfair' trade barriers. The resolution received unanimous support. For more on the views of US Government officials and US-based transnational corporations, see Hearings on the 'Globalisation of the Media', US Congress, House Committee on Energy and Commerce, Subcommittee on Telecommunications and Finance. 101st Cong., 1st sess., 15 November 1989, and the same Subcommittee's hearings, 'Television Broadcasting and the European Community', 26 July 1989.
32. Jones, quoted in 'Globalization of the Media', p.99.
33. On developments in Japan, see Yuko Nakamura, 'Direct Broadcasting by Satellite in Japan: An Overview' in Ralph Negrine (ed.), *Satellite Broadcasting, The Politics and Implications of the New Media* (London: Routledge, 1988) pp.249–68. On European developments, see the collection in Kenneth Dyson and Peter Humphreys (eds), *The Political Economy of Communications, International and European Dimensions* (London: Routledge, 1990). In the United States, direct state support for DBS is not available, although FCC regulations over domestic DBS developments have been extraordinarily lax. The official reasoning behind this approach involves the US government's role in 'encouraging DBS's market-driven development' – Personal interview with Jonathan Levy, Senior Economist at the Federal Communications Commission, Washington DC, 2 September 1992.
34. Estimate quoted in 'Hughes Looks Toward New Sky Cable Investors' in *Broadcasting* (18 February 1991) p.29.
35. See, for example, Sharon D. Moshavi, 'Time Warner Unveils 150 Channels' in *Broadcasting* (23 December 1991) p.18.
36. Another barrier involves the risks associated with the resistence of some countries to DBS 'spillover' signals. To date, an unchallenged international consensus as to the legal status of DBS cross-border broadcasts in relation to copyright and other issues remains elusive.
37. Of course television's role in the ongoing process of individual self-definition goes beyond the influence of programming content. Quite likely the very practice of the 'television experience' communicates meaning. Through the 'intimate' and isolated home location of the medium, its status as a technology 'owned' by the individual consumer, and its capacity (especially through DBS) to provide the individual with the 'freedom' to watch just about anything at any time all promote, through participation, individual consumer 'choice' as an ideal that is largely self-evident.

38. On the question of sovereignty and transborder broadcasting, see Mark J. Freiman, 'Consumer Sovereignty and National Sovereignty in Domestic and International Broadcasting Regulations' in Canadian–US Conference on Communications Policy, *Cultures in Collision* (New York: Praeger, 1984) pp.104–21.
39. From William H. Melody, from Chapter Two of this volume.
40. It should be noted that the predominant position of US film, television and advertising largely has been a symbiotic development. As Tunstall puts it, 'the success of the Anglo–Americans in marketing their media elsewhere [outside the United States] has relied heavily on their very emphasis on *marketing*. The American media just have gone further in the commercial, advertising-financed, market-oriented direction, than have the media of any other nation. The American media sell well partly because, unlike some other nations' media, they are intended for just that – to be sold'. See Jeremy Tunstall, *The Media Are American* (New York: Columbia University Press, 1977) p.131. Also see Noreene Janus, 'Advertising and the Creation of Global Markets: The Role of New Communication Technologies' in Vincent Mosco and Janet Wasko (eds), *The Critical Communications Review*, Volume II (Norwood, N.J.: Ablex, 1984) pp.57–70; and William Leiss, et al., *Social Communication and Advertising*, pp.163–76.
41. Harold A. Innis, *The Bias of Communication* (Toronto: University of Toronto Press, 1982) p.31.
42. For example, in the words of William A. Shields, Chairman of the American Film Marketing Association, national quotas on cultural products 'artificially impose and will destroy freedom of choice and inhibit the open exchange of ideas. They are anathema to the tradition of democracy'. See his testimony before Congressional Hearings on 'Television Broadcasting and the European Community', p.102.
43. From William Melody, Chapter Two.

6 International Services Liberalisation and Indian Telecommunications Policy

Stephen D. McDowell

Liberal international trade in telecommunications services is presented in two competing and sometimes inaccurate images in international political economy literature.[1] On the one hand, liberal economists depict trade in telecommunications services as a natural expansion of the liberal international trading order, an expansion which will bring benefits of accelerated economic growth to previously stagnant sectors of the economy and parts of the world stultified by monopoly state control of national service industries. While states should enter service trade agreements to exploit their comparative advantages, liberalisation is often resisted by protectionist national groups. Those drawing on dependency theory, on the other hand, point to the open 'investment' aspects of 'trade in services' agreements, and argue that the prime driving force behind institutional change is transnational corporations based in the North which wish to gain access to Third World markets. Northern market-economy states support these corporations' needs in international negotiations because such a position is seen to serve their national interests.

This chapter, however, draws on critical and historical perspectives to argue that trade in services involves more than North–South or interstate conflict and co-operation. Firstly, the historical process of international liberalisation of trade and investment in telecommunications services links a number of distinct national and international policy processes. The conventional distinctions between national state policy and international economic policy must be reconceptualised to understand adequately the characteristics and implications of liberal international trade in services.[2]

Secondly, globalisation – the deepening integration of different national markets and investment patterns – involves more than just economic patterns and interactions. It also includes new patterns of international policy formation, and new overarching images or theories. To view international negotiations merely as exhibiting conflict, co-operation and co-ordination in dealing with technical problems, or the harmonisation of national policies, is not sufficient. Globalisation goes beyond solving problems of interdependence. It also introduces the practice of assessing domestic policies by a set of international market standards. As international linkages are increasingly institutionalised or guided by organisations designed around liberal trade and investment principles, this perspective leads to an international evaluation of the role and form of national states, and the types of policies they pursue with respect to internal economic and communications policies and strategies of development. Hence, trade policy discussions cannot be undertaken in isolation from the examination of the overall role of international institutions and shifting forms of international policy rationalities.

This chapter will illustrate these historical and theoretic propositions by examining the implications of the liberalisation of international services trade and investment institutions for Indian telecommunications policies. Multilateral trade negotiations in services in the General Agreement on Tariffs and Trade (GATT) were the first in which the applicability of trade concepts to telecommunications were tested. The meaning of liberalised services exchange, both conceptually (in terms of the overall theoretic and political programme) and historically (in terms of the issues which have arisen in negotiations on services in the GATT) are examined first. The chapter then considers the implications of a liberalised telecommunications trade regime by considering the example of telecommunications policy formation in India. How will liberalised trade in telecommunications equipment and services affect the formulation of Indian telecommunications policy, and the role of the Indian state in telecommunications service provision? It is argued that trade in telecommunications services not only reduces the autonomy of national telecommunications planning, it also significantly reshapes the relationship of different international organisations and the role of the Indian state in telecommunications policy formation and service provision.[3]

LIBERALISATION OF SERVICES POLICIES: A LINKED SET OF CONCEPTS

Economic, policy and trade analyses of service activities in advanced or 'post-industrial' market economies have multiplied in the past decade. Service activities, previously viewed as non-productive (i.e. transport, health, finance, security or education) are now seen by some as the main focus of economic value creation and the central generators of new jobs.

Efforts in northern states and in several research and negotiation programmes of international organisations to liberalise rules for service production and exchange have also blossomed. In the early to middle 1980s, these measures became priorities for governments in the United States and other market-economy countries, mainly the members of the Organisation for Economic Co-operation and Development (OECD).[4]

Although the term 'liberalisation' is used to mean many things in political and economic analyses, here it refers to the greater use of market mechanisms to guide social, economic and political life (which is viewed primarily as the exchange of values and commodities among producers and consumers in different markets).[5] Applied to services, liberalisation involves analytic, policy and institutional shifts that:

1. Theoretically present the provision and consumption of services as the exchange of economic 'commodities', rather than as duties or obligations;
2. Create national and international 'markets' for the exchange of service commodities, rather than ensuring access to services either from household units, community groups or from public state agencies (i.e. security, health care, communications, transportation);
3. 'Privatise' state enterprises providing services by transferring them to the private sector;
4. Shift the forms of state regulation of service industries (often called 'deregulation');
5. Expand international trade and foreign direct investment in service activities and integrate service production on a global basis;
6. Evaluate international institutions guiding service production and exchange according to market economic and international trade disciplines as opposed to co-operative, developmental or technical co-ordination principles; and
7. Shift the forum for key international policy negotiations on service issues to those with a liberal market approach.

In sum, the various measures – commodification, privatisation, deregulation and expanded international trade and investment policies and practices – are all parts of the broader process of liberalising service institutions.[6]

This conceptualisation points out that the liberalisation of trade in services implies altered forms of state intervention and national control and guidance of service activities. The set of liberalisation policies refers not only to international trade policies. By including the formation of information commodities and markets, it confronts head-on certain notions of the role of the state in planning and national development and national patterns of state–civil society relations. National communications policies and practices are therefore connected with and are constituent parts of international practices and institutions. This linkage is consistent with the holistic theoretic framework of Robert W. Cox, which ties historical forms of state and state–civil society relations to patterns of world order.

The definition of service liberalisation also points to the importance of new policy approaches or rationalities in international organisations. The Gramscian perspective (as used by Cox, Stephen Gill and David Law, and Craig Murphy and Enrico Augelli[7]) links the fit between the relationship of material forces and forms of consciousness to the formation and operation of international institutions and world orders. Policy approaches are, in the Gramscian view, ideas and theories which have been created for the purposes and tasks of particular social groups. Sets of policy ideas are shared amongst policy researchers and economic and political groups linked to productive and destructive capabilities. Policy research and negotiations in international institutions serve to build consensus among dominant groups. However, policy approaches and negotiations also serve to give some concessions to sub-dominant groups to evoke their support of national and international political economic orders. Hence, while functionalism, interdependence and development may have been historically effective policy images and approaches for dealing with problems among northern coalitions and in North–South relations in the 1945–80 period, these understandings of international policy and practice are less relevant in a period of rapid restructuring in the global political economy. Consensus ideologies of the Fordist production and Keynesian welfare state era and moderated international liberalism are no longer appropriate.

Liberalisation also implies new roles and relations among international organisations dealing with service issues. Although some liberal analysts have argued that information services presented a 'policy void' at the national and international level, what is surprising is the significant extent of inter-state co-ordination of service activities prior to the 1980s.[8] UNESCO addressed information issues through the so-called New World Information and Communications Order. The International Telecommunication Union passed its one hundredth anniversary 15 to 20 years before negotiations on trade in telecommunications services were first proposed in the GATT in the early 1980s. INTELSAT is an interstate co-operative which has provided satellite services to all member states since the 1960s. The World Intellectual Property Organisation supported an ordered recognition of information property rights long before trade-related intellectual property measures were considered. Even among northern states, the OECD considered information and communications in a public utility framework (before rather abruptly moving to a trade conception in the early 1980s).

The historical usage of the hegemony framework in the study of international political economy also indicates that any transition in telecommunications institutions is best understood by examining the restructuring of the relationships between social forces, states and world orders, and ideas, material capabilities and institutions. Robert Cox's approach points to the organisation of production at a national and international level as being a key factor in these relationships, but it also explicitly recognises the role of ideas and (to some extent) existing institutions. Given its historical methodology and its critical purposes (to design theories which point to historical alternatives as groups with different values and purposes might define them) the approach does not attempt to specify or predict in trans-historical terms the main forces behind historical transitions. These remain matters for historical investigation.

Hence, liberalisation is a multi-faceted process that involves practical and conceptual changes in state–civil society and state–international political economy linkages, the policy rationalities used by analysts and representatives in international organisations, and the relations among international organisations dealing with communications questions. Before these implications are examined in relation to Indian telecommunications policy, it is useful to review how liberalisation concepts have been used in international negotiations. The main site for negotiations on international trade and investment in services is the General Agreement on Tariffs and Trade.

THE GATT NEGOTIATIONS AND SERVICES LIBERALISATION: COMPARING APPROACHES AND NEGOTIATING POSITIONS OF INDIA AND THE UNITED STATES

The theoretic concept of liberalisation is not entirely equivalent to how it has been applied in actual national and international policies, practices and negotiations. There has been an extensive body of work (mainly by northern analysts but also by the UNCTAD) dealing with services. In intergovernmental forums, states' interests and negotiators' positions arise from historical and political factors which shape the nature of concepts and questions, and the direction of statistical and definitional research on services issues. Although there are various developed and developing country positions in the GATT discussions and negotiations, a review of the United States and Indian proposals – these countries being the main protagonists – serves to illustrate the issues at stake in trade in services negotiations in the late 1980s and early 1990s, and better to understand the significance of the questions involved for Indian telecommunications policy.

The General Agreement on Tariffs and Trade is an international agreement (and organisation) dealing with international trade in goods and commodities. GATT membership includes mainly OECD countries in the North and market-economy developing nations in the South. One of the most important assumptions underlying negotiations is 'progressive liberalisation' – that international trade rules constantly should proceed towards reducing tariff and non-tariff barriers to trade. The expansion of international trade is seen to be the most effective policy to expand the creation of wealth for all trading partners. Another important principle guiding negotiations is 'non-discrimination', that any signatory to the agreement be given the same status as all other signatories respecting rules allowing or restricting imports. A further principle – special and differential treatment – recognises that developing countries might not be able to live up to all free trade obligations given their development objectives and needs. Hence, to a limited degree they should be given access to developed country markets even if they cannot readily reciprocate.

Services were not initially included in the GATT. In the United Nations system of international organisations, services were covered by organisations guided by principles such as co-operation, policy co-ordination, harmonisation of technical standards and development assistance. Moreover, financial transfers among states, including monetary adjustment (International Monetary Fund) and development assistance (International Bank for Reconstruction and Development or the World Bank) were covered by Bretton Woods organisations separate from the GATT.

Services were introduced on to the GATT agenda as a result of pressure from the United States. The United States and other governments responded to policy advocacy from service firms as well as long-standing economic research indicating that the United States had moved from being an industrial to a post-industrial service economy. Eventually the other northern market-economy states supported the measure. India and several other developing countries initially resisted the discussion of services in the GATT. However, due to intense pressure from the North, and after policy research in the South showed that there might be benefits for some in pursuing liberal trade in services, a compromise was reached in 1986 when, in Punte del Este, services were placed on a separate track of the GATT negotiations called the Group of Negotiations on Services (GNS). These negotiations were to take place among the GATT contracting parties following the GATT procedures.[9]

From 1987 to 1991, the GNS covered a large number of definitional issues and Indian participation became both active and intense. Along with exploring the meaning of principles such as national treatment, non-discrimination, progressive liberalisation and special and differential treatment for services trade, the GNS also dealt with the specific issues that would arise in the context of the liberalisation of six test sectors – transportation, telecommunications, financial services, construction, tourism and professional services.

The United States' position at the GATT in the 1980s reflected the general market-orientation of the American political economy and the political purposes of the Reagan and Bush administrations. The promotion of liberalised services trade included the celebration of the supposed success of market models in all aspects of economic life, including services. Deregulation of trucking and air transport, contracting the provision of public services out to the private sector and shifting forms of state intervention in and monitoring of financial industries were examples of liberal economic policies at work in the United States. The position which the United States adopted at the GATT arose from the emerging economic wisdom, promoted by significant firms in the finance and communications sectors, that market solutions provided the best option for both international trade and investment in services and for the domestic policies of other countries. Also, the strength of transnational service corporations based in the United States in selling services internationally was a crucial factor in putting services on the US government agenda.[10] The position of negotiators from the United States Trade Representative's office in the GATT multilateral trade negotiations has been consistent with this approach, despite political and intellectual challenges to the 'post-industrial' argument and its supposed liberal policy implications.[11] National and multilateral policy research on services was also combined with 'reciprocity' in bilateral trade talks and unilateral threats to cut trade ties with countries that did not respond favourably to the trade complaints of US-based companies.

In this view, the market strategy is seen to be applicable both in developed and developing economies. Markets to guide the production and exchange of services would assist development in several ways. Since intermediate services are an important input into other economic sectors, international competition in services will lower the cost of services and increase their quality. Production of services and manufactures would become more efficient and exports would be more competitive on world markets. Allowing competition by international service providers is also seen as the best way to encourage domestic providers to become more efficient and to provide more and qualitatively better services. In addition, developing countries can be competitive in international service markets by exploiting their particular comparative advantages in services (for instance, their knowledge of specific construction techniques or low labour costs for professional and skilled workers in construction projects, if labour movements are allowed).

This commitment to the expansion of service markets and the creation of world service commodities is not shared by policy analysts in the South. Post-independence Indian analysis has operated from a norm of 'development with equity', emphasising the linkage between economic growth and other social factors such as democratic participation and social equity. Development planning referred not only to capital formation, but also to measures such as political participation, social integration and quality-of-life indicators (literacy, mortality and poverty). Economic strategies included the central role of the state in development planning, with a mix of production by state enterprises and the private sector. The Nehruvian model combined the planned allocation of investment and production capacity with the use of market mechanisms so that different strategies were pursued for different sectors.[12] National self-reliance was chosen over full integration with the international economy. Many services such as banking, insurance and communications, because of their importance to national development, were either nationalised or were provided by government departments and agencies.

Indian analysts initially responded to calls for the liberalisation of national and international service production and exchange by arguing that the effectiveness of free international trade in ensuring the development of a national services sector had not been proven. Even developed countries – which now have large international service transactions – built service economies with the use of massive state spending on public services and physical infrastructure and by regulatory intervention in finance and communications, rather than by the use of markets alone. Also, outward-orientation (i.e. a focus on exporting the services and goods in which a country has a comparative advantage) has not been wholly effective for developing countries. For instance, the World Bank *Development Report 1990* notes that the poorest countries are those which have been most open and trade-dependent (especially those dependent on primary commodities trade). It adds that, 'Excluding the NICs, the other strong economies in the South are the ones which have pursued a relatively more inward-oriented strategy of development'.[13]

Secondly, Indian analysts have also made the point that integration with world markets will make the economy more vulnerable to shifts in the international economy and also reduce the role of national control over domestic service activities. Introducing a free international market in services would be socially disruptive, just as the creation of land, labour and goods markets were disruptive when first introduced in the North.[14]

Thirdly, OECD experiments with service liberalisation nationally and internationally are relatively recent. In some cases – like finance, insurance and air transport services in the United States – deregulation and privatisation have led to costly mistakes, market shakedowns and bankruptcies. These are policy 'errors' which developing countries cannot afford to make.

A United States proposal (tabled 23 October 1989) for services trade is consistent with the prescriptions of the 'idealised' liberalisation programme (possibly since many of the conceptual components of liberalisation were developed by neo-liberal economists and policy analysts based in the United States). The approach focuses on information and capital-intensive services in which the United States had relative competitive strengths. It 'seeks to open world services to the maximum extent possible. Under it, services providers would be free to locate in foreign countries and compete like local firms. They would face a fair, predictable environment for their services throughout the world'.[15]

The Indian position on trade in services has been substantially researched and refined since the initial opposition to any discussion of services in the mid-1980s. The positions put forward in two recent Indian proposals in the GATT negotiations follow from the recognition of several principles: the importance of services in development, the special needs of developing countries, the weaknesses of some service industries, and the strengths that India possesses in exporting some services.[16]

At the GATT services meeting of January 1990, India presented a proposal on the elements of a multilateral services framework.[17] The proposal laid particular emphasis on the position of developing countries. Progressive liberalisation should be 'governed by a number of principles: conformity with national policy objectives; conformity with development and technological objectives; expansion of services exports of developing countries; flexibility for developing countries to open fewer sectors or fewer types of transactions; security and other exceptions'. In late June 1990, another proposal (co-authored by India, Cameroon, China, Egypt, Kenya, Nigeria and Tanzania) was made to the GNS. The proposal covered only transactions involving the cross-border supply of services, cross-border movement of consumers and cross-border movement of factors of production, excluding permanent establishments of foreign service organisations, foreign direct investment or international immigration.[18]

The United States' and developing countries' proposals differ in a number of ways regarding how the GATT trade principles should be applied to services. The United States' proposal on national treatment is that foreign services would receive treatment no less favourable than that accorded domestic services. India proposes that national treatment be a long-term objective and not an immediate obligation. Developing countries would be free to impose entry and operating conditions on foreign service providers, and to provide preferential treatment – in the form of tax differentials, market share reservations, government procurement preferences, financial incentives, levies or surcharges on foreign service suppliers, etc. – to domestic service providers.

United States representatives argue that foreign service operators should have the right to establish an office to produce a service in the host country or otherwise to facilitate its entry from abroad. The June 1990 proposal of the developing countries stipulates that three conditions should guide trade in services: there should be a specific purpose for cross-border movement of factors of production (labour, raw material or capital); such transactions should be distinct and separable from each other; and transactions should be limited to a specific duration of time. The proposal would limit the commercial presence of the foreign company supplying services, impose requirements on foreign firms to provide training and employment in the country (as well as local content requirements) and allow developing countries to obtain access to technology and to information regarding the operations of foreign service firms. This proposal, therefore, would limit the establishment of foreign service firms in a country.

Cross-border service provision is also advocated by the United States. This refers to the right to provide a service originating in a foreign country into a host country (especially important for data services provided remotely by telecommunications).

A principle on temporary entry of service providers would be necessary to allow persons providing services to enter the territory of the host country, subject to laws governing entry. While the United States' proposal would restrict this movement to professional personnel, India is interested in allowing the temporary entry of semi-skilled and unskilled workers and the relaxation of immigration restrictions on the international flow of labour, since so many of India's service exports are labour services (such as domestics and construction workers in the Middle East and software services in the United States and Britain).

Transparency refers to the practice of making all regulations publicly known, and not using other national regulations to operate as trade restricting measures. The United States advocates a strong application of this principle, including 'the obligation to publish laws and regulations affecting services and to allow interested parties to comment on proposed regulations'. The developing countries propose a different sort of transparency. The publication of all relevant laws and regulations and administrative guidelines is also included, but it would also be imposed on service suppliers (i.e. information disclosure requirements would be placed on transnational service suppliers).

The United States supports non-discrimination – that the benefits of the agreement would be applied to all signatories (a standard arrangement for all most-favoured-nations in trade agreements). However, the proposal also includes a 'right of non-application', allowing signatories not to extend the benefits of their liberalisation to any signatory that has assumed an inadequate level of obligations under the agreement (e.g. a country which has taken reservations in an excessive number of sectors). This would limit the role in the agreement of developing countries which had taken exemptions or reserved sectors for development reasons. The Indian proposal, on the other hand, would allow exceptions to the most-favoured-nation provision to allow for the grant of preferences by the developed countries to the developing countries and for the exchange of preferential concessions among developing countries.

This brief overview provides a small sense of the complexity of the issues and the expert and technical langauge involved in negotiating changes in international policies and practices more closely to approximate to a liberal market ideal. Social implications and national concerns and interests are refracted through the prism of trade disciplines and principles, with the public 'trade' discourse divorced from the material context in which these changes in policies and practices occur. In addition, these same negotiations have also examined telecommunications as a sector to be covered by trade 'disciplines' (such as most-favoured nation and non-discrimination).[19] Next in this chapter I will relate issues in telecommunications trade discussions to the Indian policy formation process.

LIBERALISED TRADE IN SERVICES AND TELECOMMUNICATIONS POLICIES

The reader may ask how it is that debates regarding liberalised trade in services have anything to do with either international telecommunications or with the development of telecommunications in India. Liberalisation, however, confronts both national and international telecommunications policy and planning. The principles, approaches and commitments which are finally agreed upon in the GATT services negotiations and the institutional changes they imply will have important implications for the general nature of national policies involving the provision of services and for the growth of the Indian telecommunications sector.

For over one hundred years, telecommunications has been seen as a national activity with what Hudson Janisch calls an 'exchange of traffic' regime between national authorities at borders.[20] Even within national borders, the high costs of a telecommunications infrastructure meant that telecommunications was seen as a 'natural monopoly' rather than as an activity to be undertaken in competitive markets. Whatever international interaction that took place was typified by co-ordination among national telecommunications authorities over technical standards and interconnection, and the allocation of the radio spectrum through the International Telecommunication Union.[21]

In the early 1980s the perspective known as 'telecommunications for development' gained brief prominence. This research and policy programme arose as a response by international policy makers to advocacy and policy research pointing to the important role of telecommunications in overall economic development (for both developed and less-developed regions of the world). It also emerged as a way to strengthen the mandates of international telecommunications organisations, and as a means of meeting the concerns of the constituencies of developing nations within those organisations. The main elements of the programme included: increased recognition of the importance of telecommunications for overall levels of economic activity and growth; a higher priority for telecommunications in national development planning and public investment; calls for international assistance, resource and technology transfer in order to expand and improve the telecommunications network in developing countries; and the creation of concepts and indicators in order to measure and to quantify the economic value-added of interactive telecommunications (in terms of the resources saved in travel and other services).[22]

The national monopoly and international traffic exchange 'regimes' and the telecommunications for development research programme, however, were both challenged by large corporate telecommunications users in the North, and by governments that have introduced the use of more market mechanisms (deregulation, privatisation) in telecommunications service provision policies.[23]

The international trade regime which has been discussed for telecommunications services in the GATT negotiations includes a number of national and international aspects:

1. The distinction between basic telecommunications services (where national monopolies are still allowed) and enhanced services (which should increasingly move towards an international trade regime);

2. Cost-based pricing, as opposed to cross-subsidisation of different services or the use of telecommunications revenues to achieve other social and development objectives;
3. Access of all international service and equipment suppliers to national markets;
4. Access of international telecommunications users to national services on a non-discriminatory basis.[24]

In trade in service negotiations, these principles would mean the identification of 'barriers to trade' implied by certain national policies. These barriers might include the allocation of telephone lines and telecommunications services to users determined to be of national importance or the promotion of domestic value-added service providers.

The application of these trade principles would also contribute to a new hierarchy of international organisations dealing with communications issues. The post-war system of international organisations was formed in pursuit of a number of goals and principles, such as co-operation in solving common problems, development assistance, and sharing information and co-ordination in linked industries such as air transport and communication. The GATT service negotiations operate under the provision that 'attention' be paid to other international agreements and organisations dealing with services such as shipping, air traffic, intellectual property, finance and telecommunications. This attention, however, may become simply an evaluation of the wide diversity of other specialised service organisations and agreements according to market, efficiency and trade criteria alone.

IMPLICATIONS OF LIBERALISATION FOR INDIAN TELECOMMUNICATIONS POLICY

Immense challenges face Indian telecommunications. In 1989–90, India had one of the lowest telephone densities of any country in the world. Approximately 4.6 million telecommunications lines served around 850 million people. Most of these telephones were in urban areas and only 32 000 of India's 600 000 villages had *any* kind of telecommunications connection.[25] Service was widely perceived to be highly priced and of low quality. Although the waiting period for new services was significantly shortened over the 1980s, subscribers must queue up to receive new lines. While international links have improved, calls between and within Indian urban centres were often unreliable.

In the early 1990s the Indian Department of Telecommunications and Telecom Commission was in the process of creating a new telecommunications policy. Important questions for Indian policy – in the context of international service discussions – included whether to import telecommunications technology and equipment or to develop it indigenously, how to determine the role of state and private sector corporations in equipment production, whether to concentrate on expanding basic or enhanced services, what the cost structure for services should be, and the relative priority of rural or urban services.[26]

Policy questions related to the design and production of telecommunications equipment (while also relating to the question of national and international competition in equipment production) have a services trade component. Although not usually considered to be part of 'trade in services' negotiations, telecommunications equipment design and production entails extensive software engineering, and payments for intellectual property in the form of licenses to manufacture other technologies. Network service contracts may also follow from technology transfers or equipment sales. For instance, a question of central importance for the expansion of Indian telecommunications services is whether to import telecommunications switches, transmission equipment or customer premises equipment, or to develop technology and equipment indigenously. Indigenous development potentially enhances the high technology sector of the economy, prevents the outflow of financial resources and reduces technological dependence. A strategy focusing on self-reliance may, however, lead to higher costs and slower development of the network.

In India, all agree some sort of balance between the two sources of production is required. However, the exact nature of this balance is still subject to debate.[27] If a liberal trade regime is agreed upon internationally, policies promoting indigenous design and manufacture of equipment are likely to provoke trade complaints. The GATT background note on telecommunications trade, for instance, included 'access of all international service and equipment suppliers to national markets' as a principle to be considered.

Liberalisation in the imports of telecommunications services, technology and equipment is related to the overall role of the Indian state in equipment production. Up until the mid-1980s, most telecommunications switching, network and terminal equipment in India was produced by public sector undertakings (state enterprises). In the late 1980s, private production of terminal equipment increased. State enterprises were seen by some to produce high cost, low quality products which even the Department of Telecommunications has difficulty using. The state's role in providing customer premises equipment has declined to around 50 per cent from 100 per cent a few years before. While overall production capacity for telecommunications equipment was still licensed by the Department of Electronics in the late 1980s, the New Economic Policy of the summer of 1991 and the budgets of 1992 and 1993 further reduced the practice of licensing.

Enhanced telecommunications services were the focus of market and policy analysis in all parts of the world in the 1980s, and were of special concern to transnational corporate users of telecommunications. Indian business groups also asked that more enhanced telecommunications services be made available in India (in terms of types, geographic availability and quality). Business officials argue that they need high quality, low-cost telecommunications and a number of specialised services (in addition to basic voice telephony) as the infrastructure to compete in world markets and expand exports. Provision of enhanced telecom services in India has been limited by the poor condition and limited coverage of the overall Indian telecommunications network. A policy focus on meeting telecommunications demands – consistent with liberalisation – would mean that the expansion of enhanced services would become a priority. The government provided telex on demand in the 1980s and the international telecommunications carrier (VSNL) provided access to international data bases in the early 1990s. Other services such as E-mail and a domestic Public Switched Data Network were delayed.[28]

In 1990, the United States delegation to the GATT tabled complaints about India's telecommunications policies. Since enhanced services are the main concerns of large corporate users, three of the complaints addressed these issues:

1. Competition on provision of value-added telecommunications services (such as electronic mail and facsimile services) and network-based information services were not permitted;
2. The Department of Telecommunications discouraged the sale of leased lines to corporations for data transmission on certain major routes (India had installed a public packet switched network and was trying to encourage its use);
3. A [Government of India] measure prohibited United States' companies from selling the unused portion of their leased circuit capacity.[29]

Business users in India also have concerns about the pricing of local, long distance, international and enhanced services. Surplus revenues from heavy urban and international traffic areas have been used to finance service provision in high-cost areas and network expansion and modernisation. Some national business groups have advocated cost-based pricing of telecommunications services.[30] As noted above, cost-based pricing is one of the principles mentioned by the GATT background note. However, cost-based pricing for basic local calls, trunk calls and enhanced services would seriously restrict the volume of revenues available to be used to develop rural and other less-profitable services.

If 'access of international telecommunications users to national services on a non-discriminatory basis' were provided, this demand might quickly and almost fully use up a very limited resource – the telecommunications network in India. International data services would be provided over telecommunications lines within an international services trade regime. Access to the national telecommunications network would be allowed so that foreign suppliers of data services could directly reach the end user in India. However, given the low number of lines in India, international demands ultimately would conflict with development priorities.

Rural services and their relationship to development objectives are a central part of the telecommunications policy debate in India. Although the role of telecommunications in rural development received much attention in the public pronouncements of telecommunications officials (particularly the Rural Automatic Exchange developed by the Centre for the Development of Telematics), little progress has been made in expanding rural services. Some argue that development with equity will not occur until public services such as telecommunications are more accessible in rural areas. While telecommunications analysts point to the large direct and indirect economic benefits of rural telecommunications, the high costs of introducing the rural network in a large country such as India and the lack of appropriate technology and political will has restricted initiatives in this area. The emerging focus in India on providing the infrastructure for export promotion may mean that urban and specialised services (which generate quicker and larger surplus revenue streams) continue to receive priority. Pressure from relatively well-organised and articulate business groups in India and international business demands may combine with international trade principles to lower the priority of rural telecommunications.

These examples show how significant questions for India's domestic telecommunications policy are becoming increasingly linked to the formation of a new international service trade regime. Indian telecommunications policy-makers now must not only recognise International Telecommunication Union conventions, but they must also be responsive to the claims regarding trade in services made from outside the country and calls for the liberalisation of markets for telecommunications equipment and service production from within.

LIBERALISATION, THE ROLE OF THE STATE, PRODUCER AND USER GROUPS

This chapter has presented 'liberalisation' as a linked programme of national and international policy changes oriented towards the creation and greater use of market mechanisms to guide the production, distribution and exchange of service commodities. The international services order now being formed through trade and investment practices agreed upon in the GATT negotiations combines with the domestic debates over policy profoundly to affect the role of the Indian state in telecommunications.[31] International liberalisation (in the form of trade in telecommunications services) limits the ability of the Indian state and Indian corporations in designing and producing telecommunications equipment for use within the country. It places strict parameters on the emphasis that can be put on telecommunications for development planning or on meeting *needs* for rural and basic services as opposed to national and international *demands* for urban and enhanced services. A communications development strategy based upon re-directing revenues arising from heavy traffic areas and special services towards building a network in unserved areas is confronted by the new expectation or norm of cost-based pricing. Rules to promote state and national ownership in enhanced communications services are also questioned.

This chapter concentrated on the issues involved in the negotiations on trade in services in the GATT, and looked most closely at the competing proposals and concerns of the United States and Indian governments. Several questions about the broader implications of these negotiations and the direction of future developments also deserve further examination.

Firstly, the negotiations in the GATT are set in the context of a wide set of international organisations which help regulate various aspects of the global political economy. In order to understand the overall dynamics of the international political economy, the responses of the other organisations which deal with information and communications issues to the GATT's inclusion of a variety of service activities in negotiations on trade and investment should be investigated. Such an examination might consider the extent to which other bodies are now using economic and trade concepts as opposed to technical or policy co-ordination priorities as theoretic images to guide their activities. This investigation of the restructured hierarchy of international economic organisations is most usefully framed by considering Robert Cox's propositions regarding hegemonic order in the system of international organisations.

Secondly, while this chapter has provided a snapshot comparison of two initially diverging (then converging) policy approaches regarding trade in telecommunications services, the factors involved in the historical change and development of the United States and India's policies should also be examined more closely. The speed and extent to which India's telecommunications and economic policies have adopted a liberal model in the early 1990s is unprecedented. These changes cannot be understood as arising simply from negotiation processes in which India 'discovered' its interests, or simply from coercion and pressure from the North (although these factors did play a role). The discussion has also pointed to the important role of groups within India (i.e. equipment producers, large business users of telecommunications, development planners) in supporting domestic policy changes that were consistent with the liberalisation of international trade in telecommunications services. The process of liberalisation in Indian internal investment and production policies for electronics was initiated in the mid-1980s primarily because of analyses within the state. Over time, this has contributed to the strengthening of private sector producer and user groups. These groups have their own interests and their own reasons for encouraging and accelerating certain forms of liberalisation (i.e. import liberalisation and de-licensing with continued tariff and non-tariff protection). In the private Indian computer and software services industry, for example, there is support for the development of more market-oriented policies, accompanied by state support (financing, training) and opening up India's access to imports. The national private telecommunications equipment industry – now very small – wants and is likely to gain a larger role in equipment production from state-owned enterprises that in the past produced almost all of India's equipment.

The historic consensus in India on the role of the state in leading development in sectors like telecommunications is now challenged by analysts, producer groups and the consumers of these services. This is to suggest that domestic pressure for fundamental – almost structural – re-orientation may be emerging alongside the international pressures from trade bodies, development finance organisations and corporate investors. Presently, there are significant national and international forces pushing towards a single image throughout the world of what telecommunications policies should be. Information brokers in the public and private sector support the formation of intellectual property rights institutions. International users of telecommunications want a wide range of standardised services throughout the world. Investors see communications services as a new growth and expansion opportunity (such as supplying a cellular telephone service in urban areas). India's growing middle class is seen by international and national producers as a possible market for communications products and services, on the one hand, and a skilled labour force for software and information service production on the other. As argued above, understanding these changes requires careful attention to changing state–civil society complexes within India, as well as the context of broader transnational civil society within which the Indian state is placed.

Although political, economic and technical trends are not running in the same direction, several types of developments in India will likely affect the further shaping of communications policies in the 1990s. An ongoing social and political crisis in India has resulted from the breakdown of the consensus on the Nehruvian vision of a 'secular' politics and a 'socialist' state. The terms around which these conflicts will be defined and managed have yet to be determined. Liberal economic policies have been introduced in the early 1990s, despite the fact that minority governments are in power. Will the Indian state, and political and bureaucratic elites most closely connected with it, be willing or able to give up many of the state's social and developmental responsibilities? Similarly, domestic producers of communications technology and services may be torn between two images of their role with regard to the state – either to co-operate in state-led planning and investment objectives and a domestic liberalisation in return for expanding participation in Indian communications markets, *or* to move towards more direct participation in competitive international trade and investment in communications services.[32]

Notes

1. This chapter draws on research conducted in New Delhi in 1989–90. I wish to acknowledge the financial support of the Shastri Indo–Canadian Institute, and the assistance of members of the Indian Institute of Foreign Trade and the Jawaharlal Nehru University School of International Studies, institutions with which I had research affiliation while in India.
2. This conceptualisation arises from a reading of Robert W. Cox, *Production, Power and World Order: Social Forces in the Making of History* (New York: Columbia, 1987).
3. For more discussion see Jean-Pierre Vercruysse, 'Telecommunications in India: "Deregulation" versus Self-Reliance' in *Telematics and Informatics*, Vol.7, No.1 (1990) pp.109–21.
4. Among the Organisation for Economic Co-operation and Development studies on services is *Changing Market Structures in Telecommunications* (Paris: OECD, 1983).
5. My approach to the understanding of 'institutions' in political economy is centrally influenced by Karl Polanyi, *The Great Transformation: The Political and Economic Origins of Our Time* (Boston: Beacon Press, 1944). The analysis of liberalisation has also profited from reading Vincent Mosco

and Janet Wasko (eds), *The Political Economy of Information* (Madison, Wis.: University of Wisconsin Press, 1988).

6. For influential discussions of the service economy and services trade see Raymond J. Krommenacker, *World-Traded Services: The Challenge for the 1980s* (Dedham, MA: Artech House, 1984); R.K. Shelp, *Beyond Industrialization: ascendancy of the global service economy* (New York: Praeger, 1981); and Orio Giarini, *The Emerging Service Economy* (Oxford: Pergamon, 1987).
7. See Robert W. Cox, *Production, Power and World Order: Social Forces in the Making of History* (New York: Columbia, 1987); Stephen Gill and David Law, *The Global Political Economy: Perspectives, Problems and Policies* (New York: Harvester, 1988); and Enrico Augelli and Craig Murphy, *America's Quest for Supremacy and the Third World: A Gramscian Analysis* (London: Pinter Publishers, 1988).
8. See Joan Edelman Spero, 'Information: The Policy Void' in *Foreign Policy*, No. 48 (Fall 1982) pp.139–56.
9. See Gilbert R. Winham, 'The prenegotiation phase of the Uruguay Round' in *International Journal*, Vol.44, No.2 (Spring 1989) pp.280–303; and Jonathan Aronson, 'Negotiating to Launch Negotiations: Getting Trade in Services onto the GATT Agenda' (Pittsburgh: Pew Programme in Case Teaching and Writing in International Affairs, 1988). Also see Stephen D. McDowell, 'India, the LDCs, and Trade and Investment in Services' in Richard Stubbs and Geoffrey R.D. Underhill (eds), *Political Economy and the Changing Global Order* (Toronto: McClelland and Stewart, forthcoming 1994).
10. See William J. Drake and Kalypso Nicolaides, 'Ideas, interests, and institutionalisation: "trade in services" and the Uruguay Round' in *International Organisation*, Vol.46, No.1 (Winter 1992) pp.37–100. This is the most detailed historical and analytic treatment of the development of trade in services policies available in the literature. See also Stephen D. McDowell, 'Policy Research Institutes and Liberalised International Services Exchange' in Stephen Brooks and Alain-G. Gagnon (eds), *Social Scientists, Policy Communities, and the State* (New York: Praeger Publishers, forthcoming).
11. See John Zysman and Stephen Cohen, *Manufacturing Matters: The Myth of the Post-Industrial Economy* (New York: Basic Books, 1987).
12. For a good outline of development planning issues see Sukhamoy Chakravarty, *Development Planning: The Indian Experience* (Delhi: Oxford, 1987); and 'The State and Development Planning in India', *Economic and Political Weekly* (19 August 1989).
13. Sanjaya Baru, in *The Economic Times New Delhi* (23 July 1990). Also see The World Bank, *World Development Report 1990* (Washington, DC: The World Bank, 1990).
14. This point is made regarding market forces in nineteenth century Britain by Karl Polanyi, *The Great Transformation: The Political and Economic Origins of Our Time*.
15. See United States Information Service, *Economic News from the United States* (New Delhi: USIS, November 1989).
16. See UNCTAD, *Services and Development Potential: The Indian Context* (New York: United Nations, 1990), for the papers presented at UNCTAD–ICRIER

'Seminar on Role of Services in Development Process: International Experience and its Relevance to India' New Delhi, 27–29 April 1989. Also see Sumitra Chishti, 'Services and Economic Development of Developing Countries: Liberalisation of International Trade in Services and its Impact' in *The Indian Journal of Social Science*, Vol.2, No.2 (1989) pp.109–29; and Dr S.S. Saxena and Dr R.K. Pandey, *The Uruguay Round of Multilateral Trade Negotiations under GATT: An Analytic Review* (New Delhi: Indian Institute of Foreign Trade, November 1988).

17. *GATT Focus*, No.69 (March 1990).
18. *The Economic Times New Delhi* (28 June 1990).
19. Although the final content of any GATT Uruguay Round agreement is not determined as of Spring 1993, the services section will likely resemble that found in *GATT, Trade Negotiations Committee, Draft Final Act Embodying the Results of the Uruguay Round of Multilateral Trade Negotiations* (Geneva: GATT, 20 December 1991)(MTN.TNC/W/FA).
20. Author's notes from Hudson Janisch's presentation at the annual convention of the Canadian Council of International Law, Ottawa, Canada (19 October 1990). See also Hudson Janisch, 'The Canada–US Free Trade Agreement: Impact on Telecommunications' in *Telecommunications Policy* (June 1989) pp.99–103.
21. For the changes taking place in the ITU see George A. Codding and Anthony M. Rutkowski, *The International Telecommunication Union in a Changing World* (Dedham, MA: Artech House, 1982); and James G. Savage, *The Politics of International Telecommunication Regulation* (Boulder: Westview, 1989).
22. See William Pierce and Nicholas Jequier, *Telecommunications for Development* (Geneva: ITU, 1983); and International Telecommunication Union, *The Missing Link: Report of the Independent Commission for Telecommunications Development* (Geneva: ITU, December 1984). Critics might see 'telecommunications for development' as a way to deflect opinion from divisive spectrum allocation issues in the ITU and New World Information and Communication Order questions in UNESCO, as the worse type of developmentalism (state-led, high technology mega-projects which may be inappropriate to national needs and give ample opportunity for increased foreign debt and for middlemen to extract commissions) and as not addressing rural development – since the main users are likely to be urban elites and international business. See 'Introduction: Critical Perspectives on Communication and Third World Development' in Gerald Sussman and John A. Lent (eds), *Transnational Communications: Wiring the Third World* (Newbury Park, CA: Sage, 1991) pp.1–26.
23. See Peter F. Cowhey, 'The International Telecommunications Regime: The Political Roots of Regimes for High Technologies' in *International Organization*, Vol.44, No.2 (Spring 1990) pp.169–99; and Peter Robinson, Karl P. Sauvant and Vishwas P. Govitrikar (eds), *Electronic Highways for World Trade: Issues in Telecommunications and Data Services* (Boulder: Westview Press, 1989). For a critical perspective see Jill Hills, 'Telecommunications Policy: The movement towards liberalisation and privatisation' in *Telecommunications Journal*, Vol.56, No.3 (1989) pp.163–71.
24. The GATT Secretariat prepared a background note entitled, *Trade in Telecommunications Services* (Geneva: GATT, 19 May 1989) (MTN.GNS/

W/52). See also *GATT Focus*, No. 63 (July 1989) and *GATT Focus*, No.73 (August 1990) for reports on telecommunications issues arising in the Group of Negotiations on Services.

25. Chakravarthi Raghavan, 'Communication: major tool of development' in *National Herald New Delhi* (4 February 1989). These issues were discussed at a 7 October 1989 New Delhi seminar on rural telecommunications, organised by the National Telematics Forum.
26. Some of these issues are considered in Stephen D. McDowell, 'India's Telecom Policy Issues in the 1990s' in *Transnational Data and Communications Report* (June/July 1990) pp.14–18.
27. See Sunil Mani, 'Technology Acquisition and Development: Case of Telecom Switching Equipment' in *Economic and Political Weekly* (25 November 1989).
28. T.H. Chowdary, 'Indian Data Com Pent-up Demand Goes Unobserved' in *Telematics India* (November 1989).
29. United States Trade Representative's Office, 'India: Services Barriers tabled by the United States, Group of Negotiations on Services'. Unpublished (8 June 1990).
30. Federation of Indian Chambers of Commerce and Industry (FICCI), 'Business and Information Networks', background paper for Workshop on Business and Information Networks, New Delhi (16 November 1989); and National Association of Software and Services Companies (NASSCOM), 'Indian Software Industry 1990–95', background document for discussion and preparation of the perspective plan, prepared for National Software Conference 1989, New Delhi (14–15 July 1989).
31. For theoretic perspectives on the role of the state see Vincent Mosco, 'Toward a Theory of the State and Telecommunications Policy' in *Journal of Communication*, Vol.38, No.1 (Winter 1988) pp.107–24.
32. See for more discussion Peter B. Evans, 'Indian Informatics in the 1980s: The Changing Character of State Involvement' in *World Development*, Vol.20, No.1 (1992) pp.1–18.

7 Contested Terrain: Hong Kong's International Telecommunications on the Eve of 1997

Milton Mueller

In July 1997 British sovereignty over Hong Kong will end. Political control of the territory will revert to the People's Republic of China (PRC). The handover of sovereignty from a western, capitalist, colonial government to an authoritarian Chinese socialist government is a momentous historical event. Hong Kong is of great strategic significance to the regional economy, and particularly fascinating from the standpoint of a student of society. Social scientists might approach it as if they were astronomers observing some distant supernova, whose behaviour provides vital clues to the nature of the universe. By presenting such an extreme case of political transition, Hong Kong makes unusually visible the political components of economic order.

The changeover is already affecting all aspects of Hong Kong life. After signing the Joint Declaration in 1985, an apparently guilt-ridden British government launched a belated democratisation drive, complete with direct elections of Legislative Councillors and a 'Bill of Rights'. Despite assurances from China that Hong Kong will be allowed to maintain its current way of life, almost every professional-class Hong Kong resident has some kind of contingency plan to emigrate. In 1990, after residents became increasingly jittery about the implications of Tiananmen Square, the Hong Kong administration granted a limited number of Hong Kong residents British nationality in order to discourage brain drain. In 1993 a new Governor, Chris Patten, directly challenged the Chinese by proposing a political reform package that would expand the number of directly elected legislative councillors.

Most such British initiatives have been opposed or criticised by the Chinese. China may refuse to recognise the British nationality packages.

Governor Patten's political reform proposals were greeted with vitriol and threats to form a shadow government to replace entirely the standing government after 1997. As of mid-1993, the dispute over the Patten reforms is unresolved. Even before Patten became Governor, PRC officials denounced Britain's more tentative moves toward democratisation and dismissed the Legislative Council as a 'purely advisory' body. They have strongly asserted their right to consultation and/or control in matters of economic policy leading up to the 1997 handover. The Hong Kong government's proposal to build a massive new airport, for example, stirred Chinese suspicions that the British intended to bankrupt the colony before they turned it over. The Chinese demanded, and got, the right to participate in its planning and to scrutinise detailed financial information. In a similar vein, Chinese opposition has stalled British plans to privatise the colony's public broadcaster, Radio-Television Hong Kong (RTHK). The British response is that the proposal constitutes no more than an attempt to make the RTHK administration more efficient. The Chinese, on the other hand, view RTHK as a useful mouthpiece for promoting government policies and see the British attempt to privatise it as a sneaky way to deprive them of a lever for controlling the colony – a lever once enjoyed by the British.

The goal of this chapter is to analyse the impact of the 1997 transition on the telecommunications industry. I am especially concerned with the way in which China's impending control of Hong Kong will affect the trend toward telecommunications liberalisation and competition in Hong Kong. Hong Kong's telecommunications economy shares many of the technological and economic characteristics that have been associated with liberalisation in the USA, the UK, and Japan. In what follows, I will argue that Hong Kong's unique political status – its current status as a British colony and its future status as a Chinese colony – has blunted the liberalisation process. The analysis identifies Hong Kong's international telecommunications regime as the critical arena that will determine the fate of liberalisation, not only in Hong Kong itself but in China, Taiwan, and the rest of the region also.

Hong Kong is the major telecommunications hub for the Southeast Asian region and is the central point of intermediation and trade for the rapidly-growing economic area known as 'greater China'.[1] There is, however, a systemic conflict between the potentially liberal telecommunications regime represented by Hong Kong and the traditional PTT regime which still prevails in China. As the greater China economies become more integrated, both economically and politically, it will become more and more difficult for the two systems to coexist. China's attempt to preserve its monopoly PTT structure will continually constrain Hong Kong's attempt to liberalise its international regime. By the same token, Hong Kong-based attempts to liberalise international telecommunications services, to the extent that they are successful, will act to corrode the Ministry of Posts and Telecommunications (MPT) monopoly in China.

It is impossible to predict which of the 'two systems' will eventually overcome the other. One thing, however, is certain: the terrain on which the contest will be held will not be domestic telecommunications policy *per se* but trade policy. In an application of a theoretical work by Peter Cowhey[2] it is essential to recognise that international telecommunications policy in the region has been incorporated into a trade paradigm. Increasingly, the traditional telecommunications policy issues revolving around monopoly and competition, tariffing, and finance are treated not as domestic policy issues but as part of the framework of international trade negotiations, or are driven by trade competitiveness considerations. Thus the most powerful pressures for reform will come not from the economics of telecomunications technology or from domestic political constituencies, but from foreign governments and transnational corporations with an interest in penetrating markets. The obstacles to reform, on the other hand, also will be trade-motivated. Governments will resist liberalisation in order to protect domestic telecommunications service providers and the revenues they provide from foreign competition.

THE NEW TRADE PARADIGM IN TELECOMMUNICATIONS

The breakdown of national telecommunications monopolies is one of the most important communications phenomena of the late twentieth century. The traditional telecommunications administrations, whether organised as government departments linked to postal systems or as private corporations such as AT&T, were typically the largest civilian organisations and the largest employers in a country. Until very recently, monopoly control of the basic telecommunications infrastructure was an integral part of the industrialised world's political and institutional order. In Europe, telecommunications' status as a government-owned or -controlled entity dates back to the origins of postal monopolies and the formation of nation-states in the seventeenth century. Asian countries repeated this pattern as they followed the West into industrialisation. Whether one is discussing Japan, the first Asian nation to industrialise, newly industrialised economies such as Taiwan and Singapore, or still-developing nations such as China and India; whether one is looking at nations that were colonised or nations that successfully avoided Western colonisation, none have deviated much from the traditional pattern of a monopoly telephone system until very recently.

The change in the structure of the telecommunications industry can be best encapsulated by the term *liberalisation*. Liberalisation consists of four distinct but related changes in the traditional order which, taken together, move the provision of telecommunications services out of the realm of a government service and into the framework of a market economy. One is privatisation, the transfer of ownership from the public to the private sector. Another is the fostering of competition by eliminating or relaxing restrictions on new entry into various telecommunications equipment and service markets. The third, and arguably the most important, is the move away from vertical integration. In the past, one system supplied telecommunications service on an end-to-end basis, from the telephone set to switching to long distance lines. Liberalisation tends to break down this chain into its component parts, each supplied by competing, specialised firms. The fourth aspect of liberalisation is the 'rationalisation' of prices. The competitive forces unleashed by liberalisation undermine the averaged rate schemes on which telecommunications tariffs were based, compelling carriers to realign rates.

Telecommunications liberalisation is attracting a growing amount of academic attention. All too often, however, academic and industry observers accept uncritically the conventional view that the change in industry structure is a product of the intrinsic economic characteristics of telecommunications and computer *technology*. The textbook view is that natural monopoly is a product of economies of scale and scope in supply. Thus it is commonly assumed that new technologies somehow abolished these supply-side economies and thereby ended monopoly as the 'natural' form of organisation.

This argument – more often implied than explicitly stated and demonstrated – is demonstrably false. By any measure, digital switching and fibre optic transmission technologies possess enormous technical economies of scale and scope, economies which greatly exceed anything available with copper transmission and electromechanical switching. If technology and economics alone were sufficient to explain the prevailing industry structure, we should see reinforcement and extension of national monopolies, not their demise. Indeed, numerous technical manuals and public utility textbooks from the 1960s or earlier attest to the fact that the electromechanical switching systems of the past suffered from significant supply-side *diseconomies* of scale.[3] Econometric attempts in the 1970s to measure scale economies in the supply of local exchange services never established conclusive proof of their existence. Technical economies of scale in long distance services, on the other hand, were easy to demonstrate. Yet competition arose in the long distance arena and the local exchange remained a monopoly – exactly the reverse of what one would expect if the economics and technology-based explanation was correct.

This is not to suggest that new technology had nothing to do with the global trend toward telecommunications liberalisation. Explanations grounded solely in the supply-side economic characteristics of the new technology, however, provide no explanation either for the origins of the monopoly systems or for their current breakdown. Other social dimensions, particularly *political* institutions and motives, must be taken into account.

In the journal *International Organisation*, political scientist Peter Cowhey sets out a compelling view of the nature of the shift toward liberalisation.[4] In Cowhey's analysis, the origin of the monopolistic regime in telecommunications was essentially political. Politicians discovered that they could consolidate the economic and political advantages available from their control of postal systems if electronic alternatives (the telegraph and telephone) were nationalised. The average pricing and cross-subsidies of telephone monopolies were supported because rural populations tended to have disproportionate voting power. Once the political bargain creating monopolies was struck, the regime was supported by an ideology that stressed the principle that monopolies of services and equipment were the most efficient and equitable methods of providing public service, both domestically and internationally. Cowhey goes on to explain that the international telecommunications regime was organised by nation-states in ways that were deliberately designed to protect domestic carriers from the threat of external competition. Rather than allowing foreign carriers to extend facilities and services into their countries, the old regime provided international service jointly through bilateral agreements among territorial monopolies. It was, he claims, 'a multilateral framework that reinforced domestic monopolies and bilateral agreements in the global market and thereby created one of the most lucrative and technologically significant international cartels in history'.[5] Organisations such as the International Telecommunications Union and Intelsat were established to merge and institutionalise these arrangements.

The concepts of 'natural monopoly' and 'the universal network' became the basis for an 'epistemic community' justifying the regime. These economic and technical concepts were, however, after-the-fact rationalisations of what was essentially a political bargain.[6]

Cowhey assigns new telecommunications and computer technologies a catalytic role in overthrowing the legitimacy and efficacy of the old regime essentially because they altered who could 'win' and who could 'lose' in telecommunications systems. But technology only set the stage. Also important was the concentration of telecommunications needs, skills and resources among a few large users, particularly banks. These large users came into conflict with the old regime and formed a reform coalition that successfully challenged it, most forcefully in the United States. To make the scale of this challenge global, however, the reform coalition required a new set of principles and norms to supplant the old 'natural monopoly, universal network' principles.

Cowhey shows that *free trade* provided just such an alternative principle. International telecommunication was redefined as 'trade in services'.[7] This redefinition was a successful answer to the problem of finding new norms and principles on several counts: it provided a new organising principle for a competitive telecommunications regime; it allowed reformers to ally themselves with a defined and influential epistemic community (trade negotiators); it also came equipped with an alternative institutional framework (the GATT) capable of challenging telecommunications monopolies.

HONG KONG'S BURGEONING TELECOMMUNICATIONS INDUSTRY

Hong Kong is a city with a population of about six million people. While its local telecommunications services market is what one might expect from a city of that size, the volume and significance of its *international* telecommunications services sector puts it on a par with many of the world's largest nations. In 1991 Hong Kong generated 140 million international telephone calls, a volume that ranked ninth in the world, exceeding countries with much larger populations such as Italy, France, South Korea, and Mexico. Three-fourths of the revenues of the territory's telephone company are derived from international rather than local telecommunications. The growth in trade and telecommunication traffic with neighboring China is largely responsible for the size and strong growth rate of Hong Kong–international telecommunications. Hong Kong has become the financial and telecommunications hub of trade with China. In the five years since 1987, minutes of international traffic from Hong Kong to China grew by a factor of four. Hong Kong–China traffic accounts for 40 per cent of the territory's international minutes. Total outgoing international voice, fax, and data traffic grew by 26 per cent in 1991, and by 30 per cent the year before. Outgoing traffic to China is growing at 46 per cent annually

Table 7.1 Growth of International Voice Telephone Traffic, Hong Kong, 1986–91

	000 calls	*000 minutes*	*% growth min*
1986	29532	151280	–
1991	139009	612635	305

Source: Hong Kong Telecom.

Table 7.2 Proportion of outgoing minutes of telephone traffic to specific countries (%), 1986–91

Hong Kong to	*1986*	*1991*
China	19.24	39.98
USA	14.29	08.16
Japan	09.81	05.85
Taiwan	09.68	07.35
UK	08.16	04.92
All others	38.82	33.74

Source: Hong Kong Telecom.

Table 7.3 Telephone Penetration Comparison (1991)

Country	*Population (millions)*	*GNP per capita (US$)*	*Tel Lines per 1000 population*
Hong Kong	5.9	7500	513
USA	248.7	21800	540
UK	55.5	15000	400
Taiwan	20.6	7380	335
China	1151.5	1450*	10
India	866.0	1150*	5

Source: World Almanac, International Telecommunications Union (1989 data).
*International Monetary Fund, 1992.

(see Tables 7.1 and 7.2). Table 7.3 compares Hong Kong's GNP and telephone penetration ratio to other western and Asian nations.

Hong Kong's telecommunications industry is dominated by Hong Kong Telecommunications, Ltd (HKT). HKT is a holding company made up of three wholly-owned subsidiaries: the Hong Kong Telephone Company, which operates the domestic telephone network; HKT International (HKTI), which controls all international telecommunications; and HKT Communications Services Limited (CSL), which supplies deregulated products and services such as telephone sets, PBXs, cellular telephone services, and directory advertising services. Majority ownership (58.5 per cent) of the HKT group is held by the British firm Cable and Wireless plc. Cable and Wireless has owned and operated the territory's international telecommunications for many years. It acquired majority ownership of The Hong Kong Telephone Company in 1985, and in 1988 amalgamated it with Cable and Wireless (Hong Kong) to form the HKT group.

Hong Kong Telecom holds two critical monopoly licences. The first, which covers all local wireline voice telephone services, was granted for a period of 20 years in 1975. Its impending expiration in 1995 has been one of the pivotal issues in domestic telecommunications policy. The other exclusive licence covers all forms of overseas telecommunications except for broadcast signals. The international licence, which was granted in 1981 for 25 years, will not expire until 2006.

THE POLITICAL LIMITS OF LIBERALISATION

By emphasising the political roots of the regime for international telecommunications, the Cowhey framework fits the case of Hong Kong quite neatly. During the early 1980s, Hong Kong's telecommunications sector began to undergo the same liberal reforms that characterised the UK and Japan. The domestic pressures driving the liberalisation process were checked, however, by Hong Kong's unique political status as a colony.

In 1982 a reform coalition developed in the territory's telecommunications sector. Prior to 1982, the Hong Kong Telephone Company required all customers to let it and it alone install and maintain terminal equipment. This attempt to enforce the classical 'end-to-end service' arrangement began to create conflict as new and more sophisticated technologies (such as business PBXs) entered the market.[8] A number of independent companies that were trying to supply PBX equipment in competition with the telephone company were dependent on their primary competitor for installation and maintenance. This conflict of interest became even more apparent when Hong Kong Telephone created a new subsidiary called CSL to market equipment in January 1983. Another problem was that the proliferation of different kinds of equipment and models began to exceed Hong Kong Telephone's expertise and maintenance capabilities. As a result, independent manufacturers could not guarantee good service for their products.

These problems led to the formation of the Hong Kong Telecommunications Association (HKTA) by 14 major equipment vendors in July 1983. HKTA became a lobbying association promoting liberalisation in terminal equipment markets and elsewhere. Another association concerned with the needs of large users – the Hong Kong Telecommunications Users Group (HKTUG) – was established two years later. The end result of this period of upheaval was the deregulation of terminal equipment markets and the opening of cellular and paging services to competition. It should be added that Hong Kong Telephone itself supported deregulation to some extent because it wanted to free new service and equipment markets with high profit potential from government regulations that limited its profit rate to 16 per cent.

At this point the liberalisation process in Hong Kong began to take on momentum. The proliferation of equipment and service choices made corporate telecommunications the preserve of professional managers who valued choice, efficiency, and flexibility. The deregulation of cellular, paging, and terminal equipment markets was fantastically successful. Three competing cellular telephone companies were licensed. Not only did all three survive and prosper, but their rivalry quickly ushered in the era of the hand-held mobile telephone and gave Hong Kong one of the highest rates of mobile telephone penetration in the world. In 1992 a fourth operator was licensed. Hong Kong also licensed over 30 radio paging operators and, with one in seven people using pagers, attained the highest per capita subscription rate in the world.

Although Hong Kong Telecom was still dominant politically, the growth of alternative suppliers (notably cellular operator Hutchison Telecommunications) strengthened lobbying efforts for liberal policies. Government policy-makers also began to take note of the spread of competition in other countries, most notably the United Kingdom. The close political connections and movement of civil servants between colony and home country served to promote the ideas about radical deregulation that characterised Thatcher's England in the mid to late 1980s.

In 1987 Hong Kong inaugurated a broad review of telecommunications policy. Although marred by delays, contradictions and major political setbacks, the government established a general liberalisation program. In 1988 it decided to license a cable television company as part of a broad 'second network' policy to create competition in the domestic telecommunications marketplace after 1995. It established a licensing procedure (Private Non-Exclusive Telecommunications Service, or PNTS) that allowed large users to establish private networks. It liberalised the ownership of satellite dishes in 1989. In the spring of 1992, the government decided that when HKT's 1995 licence expires, domestic voice telecommunications will be opened to competition. In the course of making this change, the government will shift to a price cap form of regulation of the Hong Kong telephone company.

But Hong Kong's telecommunications marketplace is not as liberal as it appears. The primary difference between Hong Kong and other liberalising countries is that the reform process, of necessity, has been confined to *local* telecommunications. As noted before, three-fourths of the revenue and the strongest growth rates involve international rather than domestic telecommunications. Furthermore, over 40 per cent of that traffic is with neighbouring China, a traditional PTT monopoly. International telecommunications remains a monopoly of Hong Kong Telecom because of the 25-year exclusive licence granted in 1981.[9]

Cable and Wireless was granted the monopoly simultaneously with its privatisation by the British government. The privatisation of British Telecom and Cable and Wireless were both part of a Thatcher administration industrial policy aimed at creating a stronger and more globally competitive British telecommunications industry through market liberalisation and privatisation. In Britain, this generally meant more competition. Apparently, however, the UK could not resist the opportunity to use its control of Hong Kong to give a politically well-connected British firm a protected position in the international market.

The newly privatised Cable and Wireless created a new subsidiary, Cable and Wireless Hong Kong (CWHK), to operate in the territory. CWHK gave a 20 per cent ownership share to the Hong Kong Financial Secretary and received its 25-year exclusive licence in return. The conjunction of the privatisation, the licence, and the ownership role of the colonial government indicates that the deal was orchestrated at the highest policymaking levels. In this elaboration of Britain's foreign and economic policies, Hong Kong was merely a pawn – or rather, a prize.

Even after the liberalisation course had been established by the Hong Kong government, the political clout of Cable and Wireless in London dramatically slowed the domestic liberalisation process. Despite the liberal bent of key policymakers such as Secretary for Economic Services Anson Chan or Assistant Postmaster General for Telecommunications Anthony S.K. Wong, and despite vocal support for liberalisation from much of the local business-user community, it took the Hong Kong government more than five years to make the simple decision to open voice telecommunications to competition after the exclusive license expires in 1995. It has also taken Hong Kong six years to license a cable television network. This inordinate amount of time reflected not just bureaucratic sluggishness, but the Hong Kong government's fundamental lack of autonomy in relation to both Britain and China.

THE ROLE OF CHINA

The political constraints on liberalisation are complicated by the growing influence of China over Hong Kong's political economy. This influence takes the form of business investment in the telecommunications industry, but also, and perhaps more importantly, direct political intervention in Hong Kong's internal affairs. In 1991 China made it clear that it would assert and exercise its right to be consulted, if not dictate terms, on any franchises, licenses, or major government-spending programmes which spanned the 1997 transfer. The British government acquiesced on this demand, ensuring that for the next six years Hong Kong's business community is effectively subject to double sovereignty on most long-term investments.

In 1990 the territory's telecommunications industry saw a major shift in ownership which reflected the new realities. Cable and Wireless liquidated the British government's ownership share and sold 20 per cent of its stock to China's state-owned foreign investment company, CITIC–Pacific. This move gave the PRC a direct economic interest in the fate of Hong Kong Telecom. It would not necessarily be correct, however, to conclude that China will directly intervene in Hong Kong telecommunications policy to the advantage of HKT. Historically, CITIC–Pacific represents one of the most openly capitalistic arms of the Chinese state. Its former director and founder, Rong Jiren, was one of the few pre-revolution Shanghai capitalists who reached a peaceful accommodation with Deng Xiaoping and the communist government. CITIC–Pacific's primary interest in HKT may be simply that it constitutes a profitable investment. In contrast to many other issues on which the Chinese government has publicly and vehemently voiced its disagreements with British government policy, the Chinese have played almost no public role on the direction of the 1995 licence review.

Investment by China in the Hong Kong telecommunications marketplace involves seemingly contradictory forces: the investment opportunities draw quasi-governmental Chinese enterprises closer to the capitalist economic system and give them a stake in maintaining an open market. For example, two provincial branches of the Chinese Ministry of Posts and Telecommunications, the Guangdong province MPT and the Fujian Province MPT, have entered into a consortia applying for GSM cellular telephone licenses in Hong Kong's highly competitive and deregulated mobile telephone market. Indeed, the Chinese Post and Telecommunications ministry was a minor partner in the winning bid. Nevertheless, because the cable television and second telecommunication network licenses will span the 1997 handover, they must be approved by Chinese officials in the Hong Kong and Macau Affairs Department. As of this writing, the Hong Kong government is ready to license both a cable television system and alternative local networks, and several businesses are ready to construct systems, but all are awaiting the go-ahead from the Chinese. Once again, the trend toward liberalisation is stalled by Hong Kong's lack of autonomy.

Barring some major political crisis, Beijing will not take over Hong Kong's communications the way it stormed Tiananmen Square. Hong Kong Telecom and other private businesses will not be expropriated, nor will Beijing reverse deregulation and competition policies where they are already established and working (e.g. in terminal equipment and wireless services). China's political interest is defined by its need to assert its sovereignty and by its ability to maintain one-party control of the colony. As its own domestic policies indicate, China welcomes economic liberalisation but fears political liberalisation. This is why it opposes democratisation of Hong Kong's legislative council and privatisation of Radio-Television Hong Kong, but not deregulation of telecommunication services. The former, it believes, affects its ability to control and propagandise; the latter does not.

China's respect for the compatibility of its nominally socialist economy and Hong Kong's market economy under the 'one country, two systems' concept, however, does not alter the central political fact about 1997. Under the Basic Law, Hong Kong is still a colony, and thus critical elements of its political destiny will be determined by outsiders. The Joint Declaration and Basic Law do not represent unification with the mainland or liberation from colonisation, but rather a very tense changing of the guard – a northern, communist colonial power is taking the place of a western, capitalist colonial power.

Hong Kong's colonial political structure short-circuits many of the political and economic forces which have given telecommunications liberalisation an expansionist momentum in western democracies. While Beijing will not radically reverse the *fait accompli* in Hong Kong, any policies that affect the basic infrastructure, in which it now has an ownership interest, will be subject to a political veto. The PRC will not be responsive to local user demands for lower rates, more diversity, or greater flexibility. Its priorities will be shaped by the mainland government's desire for security, control, and development.

TELECOMMUNICATIONS POLICY AS TRADE POLICY

While the traditional domestic pressures will prove ineffective, trade negotiations and trade considerations provide an alternative framework for advancing or constraining liberalisation in the future. As Cowhey suggests, recasting telecommunications as a trade issue brings in new considerations and new interests.

Evidence of this emerged in January 1991 when the US State Department and the Hong Kong government concluded an agreement regarding international value-added networks (IVANs). The agreement allowed companies to resell the leased line facilities of Cable and Wireless to provide enhanced services such as fax and E-mail between the US and Hong Kong. The US–Hong Kong IVAN agreement cleared the way for AT&T and other American service providers to enter the international telecommunications market, albeit in a small way. AT&T, which perceives Cable and Wireless as a global competitive threat and resents its ability to enter the North American marketplace without restriction, lobbied strongly for this policy. Interestingly, the results of the negotiations were approved by PRC authorities before they were implemented.[10] A similar IVAN agreement was recently concluded between Hong Kong and Japan. IVANs represent a tiny portion (at most, 1 per cent) of the overall telecommunications services market. Thus, for Hong Kong, liberalisation in the crucial international sphere can occur only in those marginal areas which fall between the cracks of the monopoly licence agreement with Cable and Wireless. Significantly, global service providers, working against perceived barriers to market entry, constitute the catalysts of change.

One other possible way to advance liberalisation of telecommunications within the constraints of the monopoly license is to legalise the resale of privately leased international telecommunications circuits. Resale allows users to lease high capacity international circuits for a flat monthly fee and then resell that capacity to others. In essence, resale allows entrepreneurs to buy telecommunications capacity at wholesale rates and then sell it to the public at retail rates which may undercut those charged by the underlying monopoly carrier. In contrast to IVANs, 'simple' resale does not process or otherwise add value to the information, but simply exploits the arbitrage possibilities created by the telephone companies' pricing structures. For example, if the price of a switched telephone call is high relative to the price of voice channel capacity on a private leased circuit, a reseller might be able to lease capacity in bulk from Hong Kong Telecom and configure that capacity to sell international telephone service at a lower rate than HKT. International resellers also can set up leased-line networks that bypass high-priced international routes and enter a country in the least expensive way. For example, in Hong Kong, a company known as City Telecommunications takes telephone calls out of Hong Kong and transmits them to Canada over international toll-free lines ordered from the Canadian carrier. The network then places the call originating in Hong Kong as if it had been made out of Canada at (lower) Canadian rates. In this way City Telecommunications is able to offer a competing, discounted international telephone service without technically abrogating the HKT monopoly.

The interesting thing about resale is that it functions as the telecommunications equivalent of a free trade zone. That is, it allows users and service providers to bypass all of the 'customs barriers' built into the networks of the international PTT monopolies to protect markets and exact taxes on international communications traffic. A legalised resale regime will inexorably equalise the price of telecommunications services on both ends of an international circuit, just as free trade will tend to equalise the prices of factors of production in trading countries. It will do this because attempts to maintain 'artificial' price differences will be quickly eroded by arbitrage and bypass. Likewise, a country that adopts resale unilaterally is in a position very similar to a country that lowers its commodity trade barriers unilaterally. It is in danger of generating a trade imbalance by buying more from abroad than the other countries buy from it.[11] The redefinition of telecommunications as trade in services is not simply an arbitrary artifact of political interests – it reflects the globalisation of the industry, the increasing flexibility of digital technology, and other real changes in the material nature of international telecommunications.

For Hong Kong, a policy legalising international resale would be beneficial. It would make international rates more competitive, encourage entry and investment by new entrepreneurs, and reinforce Hong Kong's status as a hub for multinational businesses. It would also allow the territory to escape the worst effects of the HKTI monopoly without abrogating the monopoly license before 2006. A liberalised resale policy cannot be enacted, however, unless the telecommunications administrations on both sides of the international circuit agree to follow the same rules. The US, the UK, Australia, and New Zealand all appear ready to permit simple resale of international circuits. China's MPT, however, has made it clear that it is against a liberalised resale policy. Such a policy would virtually eliminate the service trade barriers protecting the revenues and monopoly position of its PTT. Resale would cut into its accounting rate revenues (which it uses as a source of foreign exchange) and possibly diminish its revenues from switched services. By unleashing the forces of arbitrage, China would subject its domestic telecommunications industry to a heavy dose of market forces. It seems most likely that China will attempt to protect its domestic telecommunications services market from the rampant foreign penetration that would occur in the wake of liberalised international resale. Here the contradiction between the 'two systems' of Hong Kong and China is manifest.

Further evidence of the importance of trade pressures in defining the international telecommunications regime can be seen from the current controversy over the renewal of China's Most Favoured Nation (MFN) status. MFN status is perceived as extremely important to China's economic development. Without it, the flow of Chinese exports to the US would be greatly restricted. Because Hong Kong is the intermediary for a large portion of China's trade with the US, an end to MFN also would have a detrimental effect on Hong Kong's economy. Under the Jackson–Vanik amendments to the US Trade Act, the MFN status of non-market economies must be reviewed annually. Thus, MFN will be a recurring issue.

In 1991 and 1992, China's MFN status came under attack by Congress because of China's human rights abuses, arms sales to Third World countries, and a growing US–China trade deficit. Given these pressures even the Bush administration (which usually followed the Kissingerian pro-China stance) attempted to use the leverage afforded by MFN renewal to pressure the PRC to improve its human rights behavior. During the 1992 Presidential campaign, Bill Clinton criticised Bush for coddling the dictators in Beijing and promised to utilise the MFN leverage more aggressively. Predictably, however, in May 1993, the Clinton administration renewed China's MFN with conditions attached to its future renewal.

One clear beneficiary of the mounting MFN pressure on China is AT&T. In February of 1993, China announced plans to purchase switching equipment from the American company (in the process snubbing the European manufacturer Alcatel) and in late March 1993, the China National Posts and Telecommunications Industry Corp., a subsidiary of the Chinese MPT, formed a joint venture with AT&T to market telephones, fax machines, and other telecommunication equipment products in China. Although Chinese officials have not explicitly acknowledged it, few observers doubt that American trade pressure contributed to the conclusion of these deals.

CONCLUSION

The 'inevitable' breakdown of public telecommunications monopolies requires certain preconditions. These include a degree of political autonomy, liberal-democratic political institutions, and a modicum of social stability whereby national security considerations diminish in importance relative to economic demands. Unless these conditions are met, the 'natural' progression of the industry toward greater pluralism, competition, and specialisation can be impeded. Politics can override the technical and economic incentives leading toward fragmentation of the single system. Political bargains were instrumental in creating the monopoly regime, and can sustain it on a domestic basis indefinitely.

Where the domestic preconditions of liberalisation are lacking, however, external trade pressures seem to be taking their place. As the markets for telecommunications services and equipment become globalised, monopoly regimes become the equivalent of trade barriers, and as such will be subject to powerful external demands for reform if, as is the case in China, a country wants to become integrated into the world trading system. The extent to which China's door is open, closed, or slightly ajar is still a matter of contention within the PRC. Post-Tiananmen, the hard-line central planners remain in control. The reform period of 1979–89, however, released about 40 per cent of China's economy from direct state control. Economic reform has strong support at the provincial level in the southern provinces. Since 1992 Deng Xiaoping has been openly promoting market-oriented trade and development, once again prodding China toward capitalist restructuring of its economy while attempting to maintain one party communist political control.

Unfortunately, the future of telecommunications liberalisation in Hong Kong depends more on how this internal contradiction in China is worked out than on Hong Kong's own internal political and economic needs. However, the more China pursues the path of market reform, the greater the importance of Hong Kong in the future development of its telecommunications industry.

On the eve of 1997, several critical areas bear watching. The most important question is how China will react to the attachment of human rights or other conditions to its MFN status. Harsh conditions (i.e. conditions which cannot be met without requiring changes that would give the impression that China is kow-towing to US pressure) could strengthen the hardliners in Beijing and lead to a reorientation away from interdependence with the western capitalist economies. Such a development would inflict serious damage on Hong Kong's trading economy and weaken its autonomy even further.

The ownership status of Hong Kong Telecom is also a vital and sensitive issue in the run-up to 1997. China's behaviour has made its intention to control the command posts of the economy by acquiring major shareholdings in strategic and infrastructural sectors abundantly clear. Less certain is whether the PRC will continue to tolerate a British firm's control of the territory's basic telecommunications infrastructure. On this front, we would do well to look for some additional manoeuvering by CITIC–Pacific, HKT, Cable and Wireless, and possibly even the Chinese MPT.

Another important political development to keep an eye on is the thawing of the relationship between Taiwan and mainland China, and the growth of political democracy within Taiwan. Taiwan, with its own military defenses and an indigenous political opposition that is not afraid to question the value of re-unification, is in a much stronger position than Hong Kong. It is attempting to privatise and liberalise its own telecommunications industry. Although progress is slow and its telecommunications service industry is overwhelmingly dominated by its PTT monopoly, it may be able to position itself as an alternative to Hong Kong as a gateway to China.

Notes

1. See Yun-Wing Sung, *The China–Hong Kong Connection: The Key to China's Open-Door Policy* (Cambridge: Cambridge University Press, 1991).
2. Peter F. Cowhey, 'The International Telecommunications Regime: the Political Roots of Regimes for High Technology' in *International Organisation*, Vol.44, No.2 (Spring 1990).
3. See for example Alfred Kahn, *The Economics of Regulation: Principles and Institutions*, Vol.2 (New York: Wiley, 1971) p.123. Also see Milton Mueller, 'The Switchboard Problem: Scale, Signaling and Organisation in the Era of Manual Telephone Switching, 1878–1898' in *Technology and Culture*, Vol.30, No.3, pp.534–60.
4. See note 2 above.
5. Ibid., p.169.
6. For a detailed exposition of how the concept of 'universal service' was redefined as a retroactive justification for telephone monopoly, see Milton Mueller, 'Universal Service in Telephone History: A Reconstruction' in *Telecommunications Policy*, Vol.17, No.5 (July 1993).
7. Some of the key literature redefining services as trade include Geza Feketekuty, *International Trade in Services: An Overview and Blueprint for Negotiations* (Cambridge, Mass.: Ballinger-AEI, 1988); Phedon Nicolaides, *Liberalising Trade in Services* (New York: Council on Foreign Relations, 1989); Jonathan F. Aronson and Peter F. Cowhey, *When Countries Talk: International Trade in Telecommunications Services* (Cambridge, Mass.: Ballinger, 1988).
8. Wong Man-Him, 'Government's Role in Information Technology: A Case Study of the Deregulation of the Hong Kong Telephone Services', M.Soc.Sci. Dissertation, University of Hong Kong, 1985. The acronym PBX stands for private branch exchange – a private telephone switching system usually located at the customer's premises.

9. For a more complete analysis of the Cable and Wireless monopoly, see Milton Mueller, *International Telecommunications in Hong Kong: the Case for Liberalisation* (Hong Kong: the Chinese University Press for the Hong Kong Centre for Economic Research, 1992).
10. Personal interview with Richard Beaird, Deputy Director, Communication and Information Policy Bureau, US State Department, 1 May 1991.
11. For a discussion of the concerns about the effects of non-reciprocal resale, see Federal Communications Commission, *Further Notice of Proposed Rulemaking*, 'In the matter of Regulation of International Accounting Rates' (CC Docket 90-337, 6 FCC Rcd 3434).

Part III
Democratic Options

8 Wired World: Communications Technology, Governance and the Democratic Uprising

Adam Jones

During the democratic uprisings of 1989–90 in Eastern Europe, Latin America, and Asia, the personal computer and the modem became indispensable tools for activists providing access to global information-exchange networks that state officials found impossible to police. Fax technology facilitated communication among groups within and outside transitional societies and provided a channel for dissemination of information when traditional networks were choked off or placed under surveillance by state authorities. Faxes and electronic mail also encouraged the proliferation of 'Urgent Action' networks established by Amnesty International and other human rights organisations. Hundreds of messages of protest could flood the offices of state authorities within the first crucial hours of detention of a rights worker or political activist. The hand-held video camera assumed sudden prominence as a means of recording state repression, evading censorship regulations, and protecting demonstrators from intervention by state forces.

The role of new technologies in the democratic uprisings of the last several years has yet to receive sustained scholarly attention. But it has been much discussed in the news media, always eager to trumpet their own political importance. An admittedly simplistic equation characterises much of this commentary: the easier the process of communication and information-sharing, the more severe the restraints on state repression, and the greater the chances for 'democracy' in its Western-style, free-market variant.

Elements of this view are presented with some rigour by Donald Chatfield, one of the few scholars to examine the ability of the latest generation of communication technology to undermine state and corporate monopoly-holders in the information sphere:

> The global information revolution influences both the flow of information and the manner in which it is analysed. Governments have historically possessed the ability to control information for propaganda purposes. More recently, corporations have also developed this capacity. The information revolution, however, shows promise to offset the control mechanisms of both governments and corporations. New patterns of information dissemination follow highly decentralised networks, rather than the old hierarchical structure. As a result, communication becomes more interactive, with less opportunity for governmental or corporate intrusion. The absence of 'noise' in new communication networks permits the flow of information with fewer ideological filters and allows citizen groups to grasp a more accurate picture of political events.[1]

This standard equation contains some important insights. But the link between communications and an alleged global 'democratic uprising' is more complex and ambiguous. In this chapter I will argue that the much-vaunted role of communications technology in sparking and sustaining democratic uprisings is solidly grounded. At the same time, however, important questions need to be raised about these technologies, particularly as they condition (and are conditioned by) an increasingly capitalist global news media.

Some of the difficulty in gauging the impact of the new technologies arises from the contested nature of terms like 'democracy' and 'authoritarianism'. Throughout the discussion that follows, I use the terms 'authoritarianism' and 'democracy' to refer to political orders that are, on the one hand, characterised by commandist, usually violently repressive, strategies of governance – with attendant censorship, limitations on freedom of association, and often direct state control of the judiciary, parliamentary structures and the news media; and, on the other hand, political orders that allow relatively free political expression and association, comparative immunity from naked state violence, and institutionalised political participation by the mass of the population.[2] Movements – successful or not – that aim to bring about the latter state of affairs I refer to as 'democratic uprisings', and their partisans as 'pro-democratic' activists or insurgents. Such uprisings are characterised, above all, by the so-called 'revival of civil society'[3] and the progressive undermining of state power and (that most nebulous of political concepts) regime 'legitimacy'.

THE MEDIA, COMMUNICATIONS TECHNOLOGY, AND DEMOCRATIC UPRISINGS: AN OVERVIEW

Television

The democratic uprisings of 1989 and 1990 were the first *electronic* uprisings, in the sense that their most galvanising images were transmitted, in real time, to the outside world. It was also true to the extent that satellite television, and other technologies, proved essential in disseminating the revolutionary message to bordering countries and beyond. The role of TV – by far the most important medium during the upheavals – forced a re-evaluation of the traditional mistrust with which political commentators have often regarded television. It became clear in 1989 that '[t]elevision knows no international borders', as Brinton phrases it,[4] and furthermore that TV's global reach could make the medium a catalyst for progressive social change – rather than (or in addition to) an agent of cultural imperialism and/or state propaganda.

In Czechoslovakia in November 1989, large crowds gathered in front of department store TV displays to watch footage of Czech soldiers beating peaceful demonstrators. The impact of the footage was magnified by the eagerness and ability of pro-democracy activists to disseminate the images as widely as possible. Film taken of the demonstrations was transferred to videotape and circulated throughout the country. In the most 'open' of state-socialist countries – particularly East Germany and Czechoslovakia – Western news reports were also accessible to the extent that an estimated 85 per cent of the East German populace watched West German TV's broadcast of state violence at demonstrations in Dresden and Leipzig.[5]

The phenomenon touched on here has become known as the 'boomerang effect'. Citizens learn about events on their doorstep from outside sources that collate and retransmit raw footage that could never be shown on state-controlled domestic media. Foreign TV and radio became 'bulletin boards of revolution'. A further, still greater political role has been regularly attributed to the news media and other information technologies. Not only did they undermine Eastern Europe-style authoritarianism, but they may actually have rendered traditional totalitarianism obsolete. Control over information has been a hallmark – perhaps the defining hallmark – of totalitarianism. But such absolute control may no longer be possible. There may be no more 'closed societies', as Mikhail Gorbachev conceded in a speech to the United Nations in 1988.

These observations call to mind some familiar themes: publicity is the enemy of repression; the power of the visual image overrides attempts at obfuscation by a repressive state apparatus. The excuses for crackdowns offered by state authorities can only be rhetorical. They lack the immediacy and attendant political significance of highly visible repression. For this reason, television (with the associated technology of satellite transmission) stands supreme as the communications technology that has most decisively altered the terms of governance in authoritarian and transitional societies. This is so not least because of TV's unparalleled impact on the Western 'home' front, where far-reaching decisions are made concerning whether support should be granted to pro-democratic insurgents elsewhere in the world.

Vietnam – the so-called 'living-room war' – brought to the small screen, with relative speed, vivid images of the fighting in Indochina and repression by South Vietnamese authorities. Isolating television's first decisive impact on a political upheaval is more difficult. But one is tempted to cite the death of ABC-TV cameraman Bill Stewart in Nicaragua in 1979. Stewart was gunned down at a Managua intersection by the Somocista National Guard. The murder was captured on film by Stewart's colleagues, and quickly transmitted by satellite to New York:

> The sequence was run on all three [US] networks that evening, CBS leading its newscast with it and ABC reserving the Nicaraguan war story for the last five minutes of its programme, devoting the time to the memory of Bill Stewart. Television stations across the United States and around the world re-ran the horrifying footage of Stewart's death several times a day for the next few days 'until you just couldn't watch it anymore', as one newsman in the States said at the time. 'People were numbed by it, and when the numbness wore off, they were outraged. Public opinion quickly registered that Somoza had to go.' The few minutes of videotape did more to injure Somoza's reputation around the world, even among conservatives, than perhaps any single incident in the decades-long family rule.[6]

The Stewart killing placed the US government, long supportive of Somoza's rule, in an increasingly uncomfortable position. Foreign policy bureaucracies – in the First World as elsewhere – are notoriously wary of on-the-spot decisions. But in societies where governments are relatively more responsive to popular sentiment, the collective emotional shock administered by certain televised images can move governments to action, however unwilling.

The most recent innovation in international broadcasting is Turner Broadcasting's Cable News Network (CNN), which rose to prominence with its live broadcasts from Tiananmen Square in 1989 and Baghdad in 1990–1. CNN has added an almost surreal dimension to the 'boomerang effect'. During the Gulf War, Baghdad residents could watch live footage of the Allied bombing of their own city courtesy of a network based in the country leading the attack. Such paradoxes aside, the main impact of CNN and its imitators for present purposes is the enormously greater space it may open for visual imagery compiled either by its own representatives or by ordinary citizens – a phenomenon to which I now turn.

'I-Witness' Video

Historically, two key factors inhibiting the flow of news and information from distant locations have been (1) lack of a sufficient communications infrastructure, and (2) insufficient resources on the part of information-gathering agencies, including diplomatic missions, intelligence services, and news media. In many respects, the former constraints have been sharply reduced, at least in urban environments.

While the second constraint is not so easily overcome, recent developments suggest it can be circumvented. In the media sphere, television news directors in the West – particularly in the US – are under increasing pressure to make overseas news-gathering cost-effective. In the face of budget constraints, the recent emphasis on viewer-friendly 'infotainment' at the expense of hard news, and the rising star of CNN, broadcasting corporations have begun to abandon the field. Foreign bureaus have closed; correspondents are 'parachuted in' to the outside world in emergencies usually when disaster or large-scale upheaval strikes. Arguably, though, these developments have been substantially offset by the large-scale increase in 'I-Witness Video'-style reporting. The new video technology potentially represents a profound decentralisation of the news- and information-gathering process. Video cameras remain expensive, but they are increasingly lightweight, sophisticated, and concealable (some are now palm-sized). In the last few years, the events these cameras have recorded – in the hands of tourists and citizens – have been used to undermine regimes, demolish official explanations, and sharply constrain the coercive apparatus of the modern state. 'In the hands of a human-rights activist', writes Colum Lynch, 'the [video] camera will likely come to be perceived as an instrument of subversion, more threatening to the state than a Molotov cocktail'.[7] Video technology has also helped to forge bonds and disseminate news among disparate activist groups and projects within and outside national boundaries:

*In Brazil, a 'national popular video association', sponsored by trade unions and the Catholic Church, oversees grassroots video productions that are fed into an emerging network for exchange and distribution. The network's adherents proclaim its power to undermine the predominance of elites, manifested in the news media and elsewhere.[8] In the Amazon, Kayapo Indians employ video cameras to record encounters with government forces, and to dissuade authorities from interfering with their demonstrations. The tactics have helped to block a huge hydroelectric dam project that would have submerged tribal homelands.

*In India, news-hungry middle-class citizens turn to 'alternative television' featuring privately-produced videotapes with features and information that put 'the sycophancy and toadyism' of heavily-censored official news sources on display.[9]

*In South Africa, Globalvision, a private TV company that distributes video cameras to residents of African townships, plans to air a regular 30-minute programme called 'Rights and Wrongs'. The programme will be devoted exclusively to video footage of human rights abuses around the world. Similar efforts are planned for Myanmar, Kurdistan, Bosnia, and other locations.

In the less developed world, the impact of the video revolution is magnified by the ease with which video technology enables foreign commentary and domestic opposition viewpoints to circulate.[10] The political impact of video has been experienced in affluent societies also. The police beating of a Los Angeles African-American man, Rodney King, prompted large-scale rioting when, in the face of seemingly incontrovertible video evidence, jurors acquitted King's assailants. A weekly newsmagazine, surveying the broader impact of video in the United States, wrote that

> ordinary people empowered with ordinary video cameras have blown the whistle on illegal garbage haulers, tuna fishermen slaughtering dolphins and stockyard workers abusing animals ... AIDS activists routinely take video cameras to demonstrations to record possible tussles with police ... Pro-life activists picketing clinics now carry cameras as well as signs ... Taken together, the camcording of America is changing the face of law enforcement, citizen action and news gathering.[11]

Computers

For decades, computer technology tended to be viewed as a threat to, rather than a catalyst for, popular aspirations. George Orwell's haunting vision of a world under electronic surveillance was bolstered by the increasing predominance of computers (and the technocratic mindset they allegedly encouraged) in the military sphere. Now World War III could be triggered by a flock of geese transformed by haywire circuits into a nuclear strike. The extent to which the massive American war effort in Vietnam came to be plotted, and its success measured, by computer projections and target selections further increased the suspicion of commentators who lamented the marginalisation of human input and ethical standards. No less grisly was the role of computer surveillance in the modes of governance employed by neo-fascist military dictatorships in Central and South America, often abetted by foreign security forces.

The technological developments of the last decade, however, compel a radical re-evaluation of the implications of computer technology for governance and, by extension, popular mobilisation. As with video, the central feature of the transformations over the past few years has been the explosive growth of accessible technologies. In this case, the personal-computer industry makes powerful hardware available to ordinary consumers in the First World and activist groups and organisations elsewhere. It has always been within the power of the nation-state to exploit technological innovations to bolster its power. Indeed, for all but the last few years of the modern era, the state – together with the transnational corporation – has held a virtual monopoly in this area. But now these forces and policies can be subverted or circumvented by activists equipped with the new generation of personal computer and modem technology. A vivid recent example is the failed coup attempt in Moscow in mid-1991. Only 2 per cent of the estimated 1.5 million personal computers in the former USSR had modems, but these proved crucial in countering the coup leaders' actions. Soviet citizens took advantage of a loose amalgam of some three dozen computer bulletin boards that gathered and disseminated information and pledges of support from major cities. Boris Yeltsin's government, ensconced in the Russian Parliament buildings, used computers to dispatch messages to aides outside the country, urging them to proceed to London and Stockholm to prepare a government-in-exile if the coup succeeded.

Chatfield likewise points to the power and prominence of computer information networks, first established in the countries of the developed North but now global in scale. One such network, the Walker Centre's China Information Centre, was founded in the wake of the Beijing Massacre of 1989. In addition to fax technology, the Centre uses computer databases and bulletin boards to disseminate information among activists around the world. Computer E-mail, meanwhile, allows the Centre to circumvent governmental controls on the transmission of documents and statements. Among the well-established bulletin boards are the worldwide

PeaceNet, which permits local members to share news of relevant events in their home areas; and EcoNet, an environmental bulletin board accessible to computer owners in seventy countries aiming to increase citizen awareness and co-ordinate political action in the environmental sphere.[12]

Fax machines

In May 1992, the world's first 'fax revolution' brought James Mancham, former leader of the Seychelles, back to his homeland after fifteen years in exile. Deposed as prime minister in 1977, Mancham had lived in the United Kingdom until, in 1989, he decided to exploit the new technology to promote opposition views among Seychellese. Mancham waged his campaign against the governing regime from a fax machine in his London home, sending seditious messages to all two hundred fax machines in the island nation. The regime's attempts to counter this information flow implicitly acknowledged that Mancham's faxed polemics had enough substance to merit a response. This in itself appears to have undermined the legitimacy of the regime, which was eventually compelled to commit itself to multi-party elections.[13]

By itself, the use of fax machines in situations of pro-democratic ferment has rarely been as prominent as in the Seychelles instance. But the role of fax technology has received extensive attention in the context of the East European uprisings and Chinese unrest of 1989–90, as well as the failed 1991 coup in the former USSR. As with computer and video, the essential development lies in a formerly prohibitive technology suddenly being made accessible to ordinary consumers in the developed world, and to a more selective – but still significant – range of groups and individuals in less developed countries.

The surge of popular protest in major Chinese cities in early 1989 brought the fax machine to the fore of opposition politics in the People's Republic. Hong Kong Chinese, and students from the PRC based at universities overseas, used the technology to send messages of support to demonstrators, along with uncensored news from the outside world. Business offices and state enterprises in the PRC were also deluged with messages. A sticker proclaiming, 'Fax Saves Lives' was plastered over Hong Kong lampposts.[14] In the wake of the imposition of martial law in May 1989, Chinese police allegedly were dispatched to stand guard over every fax machine in the country – though the surveillance likely was not so systematic.

Fax machines were equally prominent in the failed Moscow coup of 1991. One of the first measures of the coup plotters was to clamp down on the independent publications that had sprung up in the wake of Mikhail Gorbachev's policy of *glasnost*. As the censorship orders were issued by coup organisers, the *Moscow News*, one of the most outspoken independent newspapers, turned to distributing news across the city by fax. Faxes and telephones were also the main means by which Boris Yeltsin's holdouts in the Russian Parliament maintained contact with the outside world.

The wave of popular mobilisation that swept Bangkok in May 1992 similarly utilised fax (and cellular phone) technology as a means of evading state controls. Upper-class Thais – including businesspeople seeking an end to the corrupt and autocratic military regime – were among the most prominent demonstrators. Many turned their cellular phones and fax machines over to pro-democratic activists. The technology enabled demonstrators to communicate after the government disconnected their standard telephone hookups, leading some Thais to refer to the demonstrations as 'the cellular phone revolution'.[15]

INTEGRATION/INTERCONNECTION

What characterises the new technologies, above all, is their *interconnectedness*. The armoury of the pro-democratic activist is likely to be a diverse one, drawing on whatever means of communication and information dissemination are available. In this context, consider the 'Urgent Action' networks established by Amnesty International and other human-rights organisations. In the past, governments seeking to imprison, torture or 'disappear' opposition activists had a virtually insurmountable advantage – time. Most torture and extra-judicial execution occurs within hours of detention. Repressive governments have been able to exploit the *carte blanche* accorded them by the time-lag between the detention itself and news of it reaching domestic human-rights activists or the international community. But with the advent of new technologies, human-rights workers have been able to offset this temporal advantage.

The most extensive and enduring 'Urgent Action' network is that set up by Amnesty International. Ron Dart, regional development officer for the British Columbia–Yukon section of Amnesty, explains that within a couple of hours of a detention, a request or urgent action can be filed by a union or a religious community, a peasant group, or a local human-rights organisation. From there, according to Dart, news of the detention travels by phone, fax, or computer E-mail to Amnesty headquarters in London, where the Urgent Action Information Centre is based:

> London looks the situation over very quickly, and then, if they decide it's a legitimate Urgent Action, the news is usually sent to all the Urgent Action co-ordinators at the various national offices ... Each [national] section usually has a separate network, so the news is fanned out across the country by fax, telex, and mail. So on the one hand, there's the immediate response, which can usually only come from the Information Centre or head of a section; then there's people who get faxes or telexes from the section; and finally, individual members can be notified and encouraged to send their own protests.

The methods used to deliver these protest messages likewise include 'faxes, E-mail – we use the latest technological gimmicks'. Dart calls Urgent Action 'one of the essential things we [Amnesty] do. If you can catch a government while it's trying to hide its repressive acts, that embarrasses them nationally and internationally. And one of the best ways to stop human-rights violations is to embarrass the violators'.[16]

CAVEATS

The above portrait is in many respects a rosy one. Activists, inspired by images of struggle elsewhere, exploit new technologies to restrain and subvert the actions of authoritarian states. This section, however, will address some of the less straightforward implications of the communications revolution for governance and pro-democratic uprisings. The discussion follows two main lines. On one hand, I examine the significant and inevitable transforming effect of the new technology when it is applied to situations of anti-regime agitation (particularly its influence on the way the mass media transmit 'news' of this upheaval and how this, in turn, affects activist strategies). On the other hand, it is also important to consider the ingrained political and cultural biases of media organisations and personnel who often play key roles in determining what impact the new technologies (especially TV and video) will have. The first line of analysis brings several phenomena to the forefront. The most salient are, first, the possible catalytic effect of a news media-high technology presence on *state* actions, encouraging repression or severe crackdowns; and second, the way a perception *among pro-democratic activists* that 'the eyes of the world' are on them directly shapes their agendas, rhetoric, symbolism, and perceived constituencies.

What role might the news media, with their new ability to transmit real-time images of a nation's strife to a global audience, play in destabilising transitions from authoritarianism or prompting state crackdowns? The evidence is sketchy, but enough exists to caution against the simple assumption that a high-tech media presence guarantees protection for pro-democracy activists. A recent example of a state backlash spawned, in large measure, by a high-tech media presence is the Beijing events in the spring of 1989. It was, after all, the chance presence of the world's news media – particularly electronic media – during the first outbreak of pro-democracy unrest that added a sense of international drama to the protests.[17] Indeed, China scholar Orville Schell believes that 'the sheer bulk of the media [at Tiananmen] was an incitement' in itself.[18] Strategies of governance employed by Chinese authorities during the protests and occupation of Tiananmen Square displayed a profound sensitivity to the role of assembled foreign media in publicising dissent. While all governments take umbrage at unfavourable depictions of their policies, this is especially true in the less developed world. The Chinese regime, with its abiding bitterness towards foreign intervention and its decades-long campaign against 'bourgeois influences', including Western-style democratic ideology, is a particularly sensitive one. For their part, the protesters, with a sharp eye for media coverage of their activities, allied themselves explicitly and symbolically with the Western (French and American) revolutionary tradition. This seems to have bolstered the Chinese government's perceptions of creeping subversion from within. The ban on TV coverage that was eventually imposed suggested the regime's desire to implement policy free from foreign influence. As *New York Times* journalist E.J. Dionne notes, 'the Chinese Government ... had no intention of fashioning its own response to the student rebellion for the benefit of the cameras, and so it threw them [the cameras] out'.[19]

The Tiananmen events also provide an example of the influence of electronic media over pro-democratic activism. Just as US politics in the electronic age has become structured around 'sound bites' and camera angles, pro-democratic forces in authoritarian societies have grown attentive to the presence and potential of the foreign news media. But this is not just a case of exploiting a 'neutral' medium. Rather, as public-policy commentator Marvin Kalb put it, 'To some extent, the event is *changed* by having the camera trained on it ... You might say that [pro-democracy activists] are fashioning the revolution so it's coverable by the American networks'.[20] Indeed, CBS News carried footage of Chinese student activists applauding foreign cameramen and noted the predominance of English-language signs hoisted for cameras. Students prepared their banners in English and French. In symbolic terms, the most obvious act of catering to a global audience was the Goddess of Democracy statue erected at Tiananmen, with its strong resemblance to the US Statue of Liberty.

David Ignatius comments that the Chinese students 'felt themselves not just on the stage of history, as revolutionaries always do, but in its floodlights'.[21] Fundamentally and tragically, those floodlights appear to have blinded pro-democracy activists and media representatives. A general perception held that the authorities would not dare intervene while the eyes of the world were focused on Tiananmen Square:

> [T]he outside-agitator role [in the Chinese pro-democracy movement] was played by the news media, which brought the protesters onto the global electronic stage and encouraged them – and us – to imagine that we were all on the barricades of freedom together. It wasn't any conscious effort by reporters; that's just what TV does ... The Tiananmen protesters knew we were watching; indeed, they basked in our attention. But they were left alone to face the consequences.[22]

In sum, the electronic surveillance, and whatever protection it may have provided, could be stifled by government edict. And although cameras may be 'more powerful than guns' in an abstract sense, such lofty mottos are of little use against tanks and automatic weapons. Similar lessons have been learned by township residents in South Africa, Palestinians in the occupied territories, and Haitian anti-government forces, once the cameras were expelled – or as fickle foreign media personnel left for another, trendier trouble-spot.

Turning to the second line of analysis noted above, there seems little point considering the implications of the new technologies for pro-democratic activism unless one also acknowledges the political and cultural biases that determine *which uprisings* are selected for coverage, and *which elements* of them receive specific (often disproportionate) attention. Here one is able to draw on an extensive recent tradition of critical scholarship, much of it devoted to media bias in the world's reigning cultural and political hegemon, the United States. I will consider specifically political elements of the bias first, before turning to more deep-seated elements of the Western cultural tradition that also deserve elaboration.

In an article tellingly titled, 'The Media's One and Only Freedom Story', Lawrence Weschler examined news coverage of uprisings in Central and Eastern Europe during 1989–90. He contrasted this coverage with that accorded a simultaneous activist upsurge in the countries of the Southern Cone. The popular mobilisation in South America, Weschler noted, reflected rather harshly on the United States, which had installed or sustained many of the region's most repressive regimes. Weschler notes that this second 'freedom story' also called into question unrestrained free-market economic policies that had devastated South American societies but constituted a core tenet of United States' political ideology, and an implicit desideratum in US media coverage of the outside world. Not surprisingly, Weschler found the attention paid to Eastern Europe (and the role of Soviet imperialism in bolstering regimes now fallen into disfavour) was markedly greater than that paid to South American democratisation.[23]

Edward Herman and Noam Chomsky have drawn a similar comparison between the attention and righteous indignation granted 'worthy' versus 'unworthy' victims of authoritarian repression and state terror. Their well-known study contrasted media coverage of the murder of a priest, Jerzy Popieluszko – kidnapped and executed by Polish state security forces in 1984 – with the coverage of one hundred religious workers killed by US-sponsored regimes in South America, including a Salvadorean archbishop and four US nuns raped and murdered by Salvadorean soldiers in 1980. Herman and Chomsky found that

> for every media category, the coverage of the worthy victim, Popieluszko, exceeded that of the entire set of one hundred unworthy victims taken together. We suspect that the coverage of Popieluszko may have exceeded that of all the many hundreds of religious victims murdered in Latin America since World War II, as the most prominent are included in our hundred ... [W]e can also calculate the *relative* worthiness of the world's victims, as measured by the weight given them by the US mass media. The worth of the victim Popieluszko is valued at somewhere between 137 and 179 times that of a victim in the US client states; or, looking at the matter in reverse, a priest murdered in Latin America is worth less than a hundredth of a priest murdered in Poland.[24]

The issue here, then, is not so much the impact of media coverage *per se*. Any rapidly-conveyed images of popular struggles against authoritarian rule will tend to evoke a visceral sympathy for opposition forces among an international audience. The point is that the bias towards 'worthy' freedom struggles directs *how* the new communication technologies are employed: where the news anchors and video cameras head for their high-profile, on-the-spot reports; how much time and space is set aside for footage shot by participants in the struggle; and so on.

This equation of media coverage with dominant Western perspectives is not perfect. News media attention to the plight of Kurdish refugees in Turkey at the end of the Gulf War, for instance, was highly inconvenient to the Bush Administration and stimulated a retreat from original policy. But even coverage of this event suggests how the media, in reporting news from distant locales, highlight some situations and ignore others. The problem of Kurdish refugee flows to Turkey – a 'worthy' NATO member on the European periphery – was discussed at length using live satellite broadcasts. The equally massive influx of refugees to 'unworthy' Iran was a virtual non-event by comparison.

There are other, more subtle biases that influence how the new communications technologies are utilised and directed. One is the rural–urban split. The politics – and, demographically, the societies – of most developed nations are strongly biased towards the city. That is where decisions are made; where lobby groups clamour; where the political destiny and cultural identity of the nation is forged. This lean towards an urban environment is intensified by the constraints imposed by an increasingly sophisticated global communications network. While it is true that technology grows ever more mobile, it is also the case – particularly in the less developed world – that what infrastructure exists is likely to be heavily concentrated in urban areas. To the extent that political patterns and processes in these countries mirror those in the developed world, the urban arena is likely to witness the most spectacular demonstrations, and is home to most of the leading activists or intellectual figures in opposition movements. The urban bias will likely intensify as the responsibility for conveying news of foreign events to the outside world shifts further away from the roving print reporter and towards the TV correspondent – with his or her backup camera team, a satellite transmission centre in a downtown hotel, and a nearby retinue of spokespeople and 'experts' ready to be called in for the obligatory sound-bite.

The problem here is that too often the urban environment is presented as typifying the country. The events of spring 1989 in Beijing illustrate this point. For weeks, world attention was focused not just on a single city in China, but on a mile-square public space at the heart of the city. The scene at the Tiananmen tent encampment was excitingly reminiscent of 1960s student sit-ins. The Chinese student activists were colourful, quotable, and (with their English-language signs) highly photogenic. But in a country with a population that is 80 per cent rural and deeply distrustful of urban intellectuals and urban dwellers in general, serious questions must be asked about the unthinking manner in which the Tiananmen demonstrations were presented as symbolising the deeply-held aspirations of *the people*.

In this particular case, the urban bias may well have been magnified by another trait common in the West: the preference for educated, youthful, articulate, and 'qualified' representatives of pro-democratic uprisings. In Beijing, the media, with all their high-tech devices, may have missed the 'real' story. The catalytic role of student demonstrations was, in fact, quickly superseded by a mass movement composed primarily of urban workers. It was here that the bulk of state repression was eventually directed. Investigations following the June 4 attack indicate that no 'significant' violence was visited on students in Tiananmen Square. Rather, several hundred members of the workers' mass movement were massacred on approach roads in western Beijing. The workers did not conveniently congregate for the cameras in a central square. Nor were they well-outfitted with media-savvy organisers who could distribute protest bulletins in European languages or paint English-language protest signs. The fact that the majority of victims were all but ignored is not merely a point of strict historical accuracy: the misrepresentation of events in Beijing may have helped the authoritarian regime re-establish control in urban areas and reclaim its status in the international community.[25]

Similarly, during the Moscow coup attempt in August 1991, Muscovites who demonstrated their support for the Russian government – first hundreds, then thousands – were presented as the vanguard of the Soviet people (numbering more than a quarter-billion) pressing for the continuation of the free-market reform process. Reporters wandering further afield in Moscow during the coup, however, found most Muscovites registering apathy or disinterest both in the coup and the protests against it. Many stressed their desire, above all, for stability and prosperity. The rural population was barely sampled but could be expected to have displayed a similar ambivalence, given the profound economic destabilisation *perestroika* had caused in the countryside.

The discussion here has avoided the larger question of the politics of representation. The new communications technologies appear to offer a novel alternative to dominant groups' traditional representation of the less developed world (or minorities within developed societies). The means of recording life's events and popular struggles now can be placed in the hands of those actually living them. The footage can then be gathered to feed the growing market for fresh news represented by outlets like CNN. But it must not be forgotten amidst the enthusiasm that most existing exploitation of video and satellite-broadcast technologies is carried out by corporate media in developed societies, with all their particular economic interests and cultural biases – including a stereotypical view of the Third World as an eternal disaster area – and their growing preference for the ephemeral visual image at the expense of substantive investigation or a more contextual framing of events. Though the new technologies may help undermine cultural appropriation in many ways, they also reinforce it by increasing the efficiency, speed, and visual immediacy with which events in less developed and democratising societies are delivered to the outside world. The implications here for hegemony in the international system are not inconsiderable. Ian Parker, for one, views the trend towards the corporate monopolisation of international communication networks as potentially leading to 'a sharpening of regional and class divisions, as a consequence of the significantly different degree of access to available forms of information or culture between the rich and the poor'. The 'quite probable' result is a world in which '*economic* "have-nots" will increasingly become *informational* or *cultural* "have-nots"'.[26]

In closing this section, I want to consider other biases associated with the increasing predominance of *visual* media in selecting and conveying information about, and images of, the world. Western societies are founded on Enlightenment rationalism. They exhibit a preference for 'objective' evidence, most reliably apprehended by visual means as summed-up in the cliche 'seeing is believing'. Today, this manifests itself in a preference for visual imagery over the written or spoken word. As David Ignatius puts it, 'Nobody trusts anything unless he can see it with his own eyes, on TV. History happens in front of all of us, in our living rooms.'[27] As emphasised above, I acknowledge the potency of many of the visual images associated with recent pro-democratic uprisings. There are, however, inherent limits to this visual communication of information. Events can be staged or misrepresented, and the camera can become a means of propaganda *and* subversion. State agencies are adept at exploiting the public credulity that visual images evoke. Crowds pressing at polling booths to 'cast ballots for democracy' may really be there to avoid having their names turn up on death-squad lists for the 'treasonous' act of not participating.[28] The ease or likelihood of misrepresentation is increased by the now-common practice of 'parachuting' TV commentators possessing little background knowledge into emergent 'trouble-spots'. Moreover, the emphasis on visual imagery leads inevitably to a focus on finite, dramatic events rather than broader political contexts or longer-term developments. A growing body of scholarship demonstrates the leaning of the Western mass media towards stories about coups, earthquakes and hostages.[29] Political transformations often lack this kind of ready visual imagery. What images exist may be seized upon and blown out of proportion by the custodians of visual media, presenting a distorted picture of the scale or context of events.

Two other biases seem evident in the transformation of television news into a real-time phenomenon, best exemplified by the rise of CNN. The first appears to require a revision of Marshall McLuhan's celebrated dictum: the medium is no longer just the *message*, it is also becoming the *story*. Camera crews take pictures of each other taking pictures of events. This practice, prevalent in coverage of the Gulf War, reached new depths of absurdity with the US landing in Somalia. Media representatives appeared to outnumber troops, and most pictures were framed to show the banks of media in close proximity to the soldiers. The possibility of substantive investigation to explore pressing political realities grows ever more evanescent. Secondly, and more subtly, satellite technology has spawned the phenomenon of 'instantaneous journalism'.[30] The role of the reporter as interpreter, mediator and framer of the visual image becomes ever more peripheral. The media cease to mediate between the raw image and the distant viewer. This results not in a 'neutral' transmission of events, but in communications that are more open to distortion and careless misrepresentation. It also may result in self-interested manipulation by state authorities, with their considerable resources, their ability to provide site access, and their stock of well-groomed, camera-friendly official spokespeople.

CONCLUSION

This examination of communication technology's impact on authoritarian governance and pro-democratic activism provides much in the way of support for the standard view of technology as potentially a liberating force. The new communication technologies have, on the whole, been a boon to pro-democratic activism the world over. It also is clear, however, that these technologies are by no means discrete phenomena operating independently of the agents who utilise them. Rather, they flourish in an increasingly interconnected world that, for all the decentralisation and 'democratisation' of information the new technologies permit, is still dominated by Western state interests and transnational corporations. These serve to bolster the power of the West's cultural and political models, motifs, and values. Corporate media and the 'culture industry' – motion pictures, music, television, and so on – have seized on the new technologies in a manner that must, unfortunately, mitigate any comfortable equation of technology with democracy.

The positive impact of the new technologies is most apparent at the grassroots level. The communications revolution has fundamentally transformed the strategies and potential of pro-democracy activism, and has placed powerful constraints on the ability of authoritarian forces to suppress anti-regime organisation and mobilisation. In this sense, the terms of governance have been recast, a phenomenon which is also evident (albeit on a lesser scale) in the developed world. But when we view state–society relations in the broader context of patterns of global hegemony, technology's impact is more ambiguous – even ominous. The new porousness of borders does not simply permit the influx of neutral, disinterested philosophical influences from the outside world. Rather, it intensifies the penetration of ideologies, models, paradigms, and strategies that may or may not be appropriate to a given society's popular aspirations. Of course, this argument relies on subjective notions of the 'integrity' and 'autonomy' of democratic processes. But certain effects of the new technologies can be isolated with greater confidence. In particular, the extent to which these technologies may *exacerbate* tensions between a regime and its opponents; *encourage* unrealistic expectations on the part of activists; and *provoke* state repression, partly owing to leaders' perceptions that their sovereignty and legitimacy is unfairly undermined by the introduction of outside representatives and ideologies.

The most important variable here appears to be the presence of corporate news media – riding the wave of new communication technologies, but limited by their own institutionalised conceptions of the less developed world, and by an obsession for Western models of democracy. Yet even this bleak appraisal has a positive dimension. It highlights the basic role of human agency in exploiting the new technologies. The challenge for pro-democratic activists and their sympathisers in the international community is to maximise the liberating potential of the new technologies, while working to constrain the manipulations of those for whom 'democratisation' is treason – or just another sound-bite between commercials.

Notes

1. Donald Chatfield, 'The Information Revolution and the Shaping of a Democratic Global Order' in Neal Riemer (ed.), *New Thinking and Developments in International Politics: Opportunities and Dangers* (Lanham: University Press of America, 1991) p.159.
2. These definitions ignore the contentious issue of economic democracy, an inevitable shortfall given the limited space available. I do, however, consider massive disparities in resource distribution to be inimical to a democratic order. This perspective is perhaps implicit in my later discussion of the new communications technologies as potentially redressing such inequalities in the information sphere, and also in the concerns I raise over the increasing monopolisation of the mass media by First World (i.e. disproportionately privileged) members of the global community.

   Only when significant transformations in economic power occur does a democratic uprising become a 'revolution'. I avoid this latter term as much as possible since much of my evidence is drawn from instances where democratic uprisings have clearly taken place, but where the 'revolutionary' balance-sheet is a good deal more uncertain. In any case, no revolution takes place without an uprising of some kind, even if relatively few uprisings can spark the deeper transformations that revolution entails. And it seems to me that the uprisings themselves are worthy of study (and usually support), even if they succeed only in mitigating the worst aspects of authoritarian governance.

3. For discussion, see Guillermo O'Donnell, Philippe Schmitter, and Laurence Whitehead, *Transitions from Authoritarian Rule: Tentative Conclusions* (Baltimore: Johns Hopkins University Press, 1986) ch.5.
4. William M. Brinton, 'The Role of Media in a Telerevolution' in William M. Brinton and Alan Rinzler (eds), *Without Force or Lies: Voices from the Revolution of Central Europe in 1989–90* (San Francisco: Mercury House, 1990) p.468.
5. Brinton, 'The Role of Media', p.460.
6. Bernard Diederich, *Somoza and the Legacy of US Involvement in Central America* (New York: E.P. Dutton, 1981) p.271.
7. Colum Lynch, 'Recording Repression: The video is mightier than the sword' in *The Globe and Mail* (14 November 1992) p.43.
8. Joseph D. Straubhaar, 'Television and Video in the Transition from Military to Civilian Rule in Brazil' in *Latin American Research Review*, Vol.24, No.1 (1989) pp.150-1.
9. Barbara Crossette, 'In India, News Videotapes Fill a Void' in *The New York Times* (2 January 1991) p.A6. The quote is from Madhu Trehan, creator and an anchor of one of the video news programmes, 'Newstrack'.
10. Douglas A. Boyd, Joseph D. Straubhaar and John A. Lent, *Videocassette Recorders in the Third World* (New York and London: Longman, 1989).
11. 'Video Vigilantes' in *Newsweek* (22 July 1991) p.43.
12. Chatfield, 'The Information Revolution', p.146 and p.149.
13. 'Fax revolution helps bring back democracy' in *The Guardian* (10 May 1992) p.B5.
14. See 'China tries to pull plug on fax machines – and the outside' in *The Gazette* (12 June 1989) p.B1. The report does not specify whether the stickers' slogan appeared in English or Chinese.
15. Philip Shenon, 'Mobile Phones Primed, Affluent Thais Join Fray' in *The New York Times* (20 May 1992) A10. Unlike fax machines, however, which utilise existing telephone lines, cellular technology relies on an infrastructure that must be created from scratch. Its role has thus been strictly limited in the most recent round of pro-democratic uprisings, and limited further to relatively affluent urban areas.
16. Personal interview with Ron Dart, Vancouver, 11 December 1992.
17. Recall that the news media were in Beijing to cover not anti-communist popular rumblings, but Mikhail Gorbachev's first state visit to China.
18. Quoted in David Ignatius, 'Media were actors in Beijing tragedy' in *The Washington Post* (2 August 1989) p.B3.
19. E.J. Dionne Jr., 'TV Steps Into the Fray, and Alters It' in *The New York Times* (21 May 1989) p.A18.
20. Quoted in Dionne, Jr., 'TV Steps into the Fray' (emphasis added).
21. Ignatius, 'Media were actors'.
22. Ibid.
23. Lawrence Weschler, 'The Media's One and Only Freedom Story' in *Columbia Journalism Review* (March/April 1990) pp.25–31.
24. Edward S. Herman and Noam Chomsky, *Manufacturing Consent: The Political Economy of the Mass Media* (New York: Pantheon, 1988) p.39.

25. Robin Munro, 'Who Died in Beijing, and Why' in *The Nation* (11 June 1990) p.811.
26. Ian Parker, 'Economic Dimensions of 21st-Century Canadian Cultural Strategy' in Parker et al. (eds), *The Strategy of Canadian Culture in the 21st Century* (Toronto: TopCat Communications, 1988) p.224. Similar concerns have been raised over the last two decades by Third World proponents of a New World Information Order. See, for example, D.R. Mankekar, *Media and the Third World* (New Delhi: Indian Institute of Mass Communication, 1979).
27. Ignatius, 'Media were actors'.
28. The tradition of obfuscation and misrepresentation here is a long one. For an overview, see Edward S. Herman and Frank Brodhead, *Demonstration Elections: US Staged Elections in the Dominican Republic, Vietnam, and El Salvador* (Boston: South End Press, 1984) esp. 'The Role of the Mass Media in a Demonstration Election', pp.153–80.
29. See Robert A. Hackett, 'Coups, Earthquakes and Hostages? Foreign News on Canadian Television' in *Canadian Journal of Political Science*, Vol.22, No.4 (December 1989) pp.811–25.
30. The phrase is Michael Kamen's, quoted in Barbie Zelizer, 'CNN, the Gulf War, and Journalistic Practice' in *Journal of Communication*, Vol.42, No.1 (Winter 1992) pp.66–79.

9 Developments in Communication and Democracy: The Contribution of Research

James D. Halloran

> We need the knowledge that only research can provide before we can develop adequate communication policies.[1]

THE NATURE AND PURPOSE OF COMMUNICATION RESEARCH – IN WHOSE INTEREST?

The above quotation is taken from a UNESCO Report of over 20 years ago which, amongst other things, set out proposals for an international programme of communication research. Although it might have been made clearer in the Report that knowledge, although a necessary condition, is not in itself sufficient for appropriate political action and desired outcome, the report, stemming from an earlier working paper,[2] clearly marks a shift in thinking about the nature and purpose of international communication research.

In this approach we do have, arguably for the first time, a set of proposals in an international setting, clearly indicating that research should be a public concern, policy-oriented and firmly tied to democratic developmental requirements. Gradually we begin to hear about 'meeting the communication needs of society'.

I am suggesting that this development represented a clear shift in research orientation at the international level, for hitherto communication research had not been conspicuous for its interest in democratic issues. It had been geared more to market, commodity and economic growth interests than societal concerns, and this applied to much of the early UNESCO-related research with regard to the Third World, as well as to the commercial needs of the First World.

The base-line of this development may be further illustrated at a more specific and topical level by referring to the proceedings of another meeting around the same time, this one organised by the International Broadcast Institute (now the IIC) on The New Communication Technology and its Social Implications, and attended by policy-makers, media practitioners and academics from all over the world.[3]

Consensus was not the hallmark of this international gathering, but many democratically related issues were raised, including the need to develop some form of social accountability, and that an approach should be adopted that would enable the probable costs and benefits of any act of innovation to be assessed before the innovatory decisions were made. Every development could have a 'fall-out' cost in a range of related areas; therefore we needed to study this double-barrelled character of technological innovation, for both the positive and negative effects often occurred at the same time, and in virtue of each other. It was also suggested by some that we were essentially dealing with a public matter – something which fell within the sphere of public as distinct from private interests and that the relevant institutions and the decision-making process should take this into account. This view maintained that decisions in this vital social area ought to be taken away from the free-for-all of the market place.

That these and many other questions that were raised could not be answered solely by social scientific research, because value and policy considerations were involved, was clearly recognised. It was also recognised that without the data from research very little progress could be made, and it was regretted then, as it still is today, that there seemed to be no proper means for formulating policy on new developments on the basis of adequate research and reflection.

In most cases the appropriate research had not been carried out mainly because the support for such research had not been forthcoming. But there were those who argued that even if the support had been forthcoming, the ideas were not there, or at least if they were, no-one had taken the trouble to operationalise them by formulating appropriate designs.

It was also pointed out that lack of information on vital issues had rarely been allowed to interfere with policy formulation and decision-making and even when data were available there had been many instances where the reports had been quietly shelved if they had not been judged supportive. Moreover, it was recognised by some that although we might call for more research and assert that as much research as possible should be carried out before decisions were made, it would be naive and unwise to forget that for many people, particularly those with vested interests and something to lose, ignorance could be functional and new factual information could often be a source of embarrassment.

Research may never be contemplated. It may be proposed and turned down; it may be sponsored with strings attached; it may be carried out simply to give the impression of concern; its results may be ignored, rejected, selectively used, distorted, accepted, and sometimes acted on. What eventually happens depends on many factors, and it is these factors that will be examined later.

When we turn to the present day and look through the research journals we find that more or less the same general concerns are being expressed as in the earlier years. Attention is still drawn to both 'negative' and 'positive' possibilities, and the main thrust is still very much in terms of what we *ought* to be doing in order to avoid disasters and protect the public interest.

Today there are many more publications and relevant articles than there were twenty years ago, and many more voices contribute to the debate. A reading of three relatively recent publications,[4] picked more or less at random from the growing pile, provides a good idea of the nature and scope of the current debate, highlights the main concerns, and even points to a few hopes for the future.

The language has changed somewhat and new terms have been introduced in order to deal with the emerging situation. In addition, the debate has widened as more specific problem-areas have been forecast (e.g. centralisation of power; the by-passing of library-centred information systems; the buying and selling of information; the problems of the surveillance society; and the decline in public service broadcasting) and as the wider implications at political, economic, social, cultural, international and individual levels of 'the dramatic and pervasive changes' in the communication system have been identified or hypothesised.

Some still talk in terms of 'the myth of the information revolution'[5] in the sense that the lives of ordinary men and women have not yet been radically changed. But the pace is definitely quickening, and although we may have waited for it much longer than some anticipated a quarter of a century ago, there are not many today who doubt that, even if it is not already with us (obviously national differences are relevant here), we shall soon encounter an information society in which, according to Bill Melody, the most distinguishing characteristics will be 'the increasing dependence of institutions and people on particular kinds of I[nformation] and C[ommunication] in order to function effectively in their economic, political, social and cultural activities'.[6] In a sense this represents the reinforcement of a dependency situation in which information/knowledge will be more central than ever before.

Those who initiated these debates and voiced their concerns in the late sixties and early seventies[7] were fully aware of the folly of focusing on the media or on communication technology to the exclusion or other contextual factors. Economic, political and social trends in recent years (at least in some countries) have shown how right they were. Privatisation, denationalisation, the erosion of the spheres of activity of the public sector (not just with regard to communication, but in education, health and elsewhere), the overriding importance of market forces and commercial considerations (even in the public arena), deregulation at the economic level, accompanied by more regulation at the political level, and the growth of transnational operations have certainly done nothing to calm the fears of those who have warned us about the dangers and disadvantages of the information society if introduced in certain ways.

It is quite legitimate to have these fears without adopting a deterministic stance. One might accept, with Gillespie and Robins, that 'Technology has an inherent "bias", for it can never be neutral or independent of society's broader social and political biases. At the same time, however, its potency makes it invariably the site and stake of struggle – the outcome of which is never preordained.'[8]

There is no evidence to support the inevitability of special social/cultural outcomes stemming from the brave new world of the information society. We need to recognise that we are not dealing with a one-way process; the situation is certainly not characterised by 'the unopposed invasion and conquest of the passive'. People are not all passive accepters. Admittedly, they may welcome, accept or collude in some cases, but in others they may ignore, select, reshape, redirect, adapt and on occasions even completely reject.

Moreover, as emphasised earlier, all these processes must be seen in the light of the political and economic climates and decisions which influence the way technology is introduced, and which govern its organisation and *modus operandi*. Nothing is inevitable and, as has been dramatically demonstrated on more than one occasion, the people are not without power. In a way this one-sided approach is a further reflection of the technological/media centredness that has dominated so much of our thinking and research in the past. It is time we reversed the process and turned the equation around. We should not take social, political, cultural outcomes for granted, or regard them as the automatic concomitants of some technological/media innovation or arrangement.

Accepting all this, however, does not necessarily improve the position. In fact, it is all too evident that the ways in which all these factors are interacting at the present time suggest that the gaps between rich and poor, at both national and international levels, are likely to become greater, and in the future there will be fewer opportunities for certain underprivileged groups in society to have access to information (which is the *sine qua non* of effective functioning in a participatory democracy), and that, in general, we seem to be incapable of, and/or unwilling to manage the developments in a way conducive to the public interest.

But intervention and action should not be ruled out. The Congress of the United States Office of Technological Assessment, in pointing to the part that government might play in the realm of communication, drew attention to both the 'opportunities and constraints presented by the new communication technologies'. Opportunities and difficulties may be viewed in relation to four sets of interests. These are Business, The Democratic Process, The Production of Culture, and The Individual. These interests may, of course, conflict with each other; success in one area making failure in another well-nigh inevitable. But it is worth noting that it was recognised that, before advantage can be taken of the opportunities in the three non-business arenas, some form of 'public intervention' will clearly be necessary. Moreover, it is also recognised that such intervention cannot be narrowly confined to 'information and communication', for people need to be adequately equipped with the skills and competences which would enable them to take advantage of whatever is provided. Equality of access is obviously important, but even in the unlikely event of this taking place, we need to remember that provision is not the same as use, and that information cannot be equated with communication.

In his review of the overall situation, Bill Melody concentrates on the public interest, and on the policies that would meet public interest requirements and other democratic criteria in the anticipated future.[9] He also suggests ways in which we might give more attention to those public interest implications, hitherto neglected, and emphasises the importance of attempting to make good our lack of knowledge by an increased research effort. We need research which would provide an overall systematic analysis of the long term implications of technological developments and associated institutional changes.

This is fine, but unfortunately it differs little from what was being presented as urgent some twenty years ago. If we are still calling for such action today – and rightly so – does this mean that very little has been done over the past two decades? If this is indeed the case (particularly with regard to the impact of research on policy) – and I think that this can be argued – then clearly we must ask why. In passing, we might also ask ourselves to what degree, if at all, have communication policies been influenced by research in cases other than those where the research simply reinforced the established system?

In March 1972 an International Symposium on Communication was held at the Annenberg School of Communications at the University of Pennysivania, and the paper I prepared for one of the panels had the title 'What do we need to know? Are we going to be able to find out?'.[10] That question worries me as much today as it worried me then – perhaps even more so. That it is still necessary for us to call for an increased research effort in order to answer the questions that have been on our agenda for so long is because we have not been able and/or have not been allowed to answer them.

I think Bill Melody is absolutely correct to stress the need for an overall systematic analysis of the long-term implications of technological developments, even if it has been said many times before. But what I think is more open to question is Melody's belief that the academic research community (even after remedying the shortcomings which Melody recognises) is ideally placed to undertake this task. In certain circumstances I think that the academic research community, if it got its act together and put its house in order, might be capable of this task, but I think that there are some indications, historically and currently, that suggest that such circumstances may never exist.

RESEARCH DEVELOPMENTS : WHAT PROGRESS?

It is, then, to certain selected aspects of the history and nature of communication research that I shall now turn, if only to see if there is any ground for these doubts and fears. I do this not in any pessimistic or defeatist mood, nor in any way to deny our considerable achievements, but primarily in an attempt to see if what we can learn from the past might help us to identify the obstacles, internal and external, that will have to be overcome before we can achieve the aims which I am sure we all share.

The first point I wish to make has to do with the aforementioned UNESCO Montreal Conference – the research and critical thought, particularly about international communication, that stemmed from it; the development of a well-financed and well-orchestrated counter-attack against critical research (but particularly against UNESCO's involvement) by the international communication establishment; and an eventual retreat from the critical arena.

My interpretation of the situation is that the main hope of quite a number of researchers at the time of the Montreal meeting in 1969 was that some form of critical approach – not homogeneous, not representing any given ideological position, but diverse and pluralistic – would take over from the conventional research which, until then, had characterised both the field in general and UNESCO's research policies and programmes in particular.

This earlier type of research had far-reaching policy implications. For example, as far as communication development in the Third World was concerned, implicit in these models of research (but rarely explicitly stated) was the idea that development in the Third World should be measured in terms of the adoption and assimilation of Western technology and culture. The main emphasis of the work was on increasing efficiency within an accepted and unquestioned value framework. In general, prior to Montreal, many of the projects sponsored by UNESCO (deficient in theories, models, concepts and methods) tended to legitimate and reinforce the existing system and the established order, and in the Third World it tended to strengthen economic and cultural dependence rather than promote independence. It would appear that it was this conventional research tradition and its advocates which were approved, and the break from which was clearly deplored by those involved in the counter-attack who represented the international communication establishment.

It was only after Montreal that the questions raised in research about mass communication became more relevant, challenging and provocative at both intellectual and political levels. It was this that caused all the trouble. Those who control international media and communication operations, and those who serve them, tend not to be interested in criticism, challenge, stimulation or alternatives.

Immediately after Montreal, UNESCO-related research (at least in certain areas) became more critical, and eventually this led to the production of research reports that, for the first time, spelled out the true nature of the flow of international information, media materials, etc; described the influence on this flow of historical, economic and political factors; and pointed to the inevitable outcomes of the operation of the free-flow doctrine in a world where national and regional communication resources and capabilities were so unbalanced.

It was this exposure of the implication of free-flow and other related revelations (coming from several different sources) that provoked the attack on UNESCO's research policy from those who, until that time, had shown little interest in research other than that which reinforced their position. Prior to this there had been very few, if any, systematic or sustained challenges from research to any established communication structures or policies. The function of this conventional research had been to support the status quo. What some saw at the time as obvious inadequacies had been hidden for years. Clearly, the opening up of these issues was regretted by many of those who spoke and wrote in terms of 'freedom'. Those who voiced disquiet by means of their research were attacked by those who so obviously benefitted from the silence of the past.

One of the outcomes of all this (and, of course, there were far more important international, political and economic outcomes) was pressure to shift research away from such questions as 'the right to communicate' to 'more concrete problems', although these were never clearly defined. This, rationalised in terms of the necessary depoliticisation of research, was obviously an attempt to put the clock back to the days when the function of research was to serve the system as it was, and not to question, challenge or attempt to change it.

It is worth noting that it was stated at the time that UNESCO-related research had become increasingly politicised. There was no recognition that research had always been politicised in one way or another. The complaints about increased politicisation were usually related to this comforting blindness, and the related failure to distinguish between the latent and the overt. The complaints and criticisms were based on the belief which regards conventional research as value free and truly scientific and any research which departs from this 'scientific approach' as being 'unscientific', 'philosophical', 'qualitative' and 'politically motivated'.

In making these points it is not being suggested that nothing critical emerged from UNESCO in the decade that followed and, of course, critical thought flourished in areas other than mass communication research. The main point is that this is an example of a well-organised and well-financed set of negative and hostile reactions to a research effort which was clearly geared to the public interest. As was clearly stated at the time, research which concerned itself with societal or public interest objectives, and which sought to inform policy-makers, was not what was required. We must now ask if the situation is any different today.

Although there are differences from country to country, even from institution to institution, I think lessons similar to those just outlined may also be learned from attempts to develop co-operative research exercises between the academic research community on the one hand, and media institutions, media practitioners and policy-makers on the other. I am, of course, referring to independent, critical research (not to the research which simply aims to serve the institution or industry on its own terms), and in particular to that kind of research which may require the media institution or communication industry to provide access and facilities, not to mention financial support.

Not surprisingly, media practitioners and policy-makers, although stating that they would welcome help from research, tend to be selective in their reactions to, and use of, research results. Amongst other things, they prefer researchers to deal with problems they have identified and defined themselves, and they rarely welcome 'external' definitions from independent researchers which suggest that there may be other problems which are more important, both to society and to communication. Understandably, they do not welcome research which challenges their basic values and deeply held beliefs, or which questions their well-established and accepted professional ways of doing things. Why should they? It is as well to remember that the two groups (researchers and those who work in the media/communication industries, and policy-makers) may have no common points of reference.

However, we need to recognise that if co-operation has been lacking on vital issues – and it has – then this need not be entirely the fault of those working in the media and the communication industries. Research obviously has an important part to play, but it is not without its limitations and shortcomings. Whilst we may all accept that research is essential in order to provide the base for informed policy-making, it would be misleading and very unwise for researchers to create false expectations, and to suggest that successful formulae and clear answers could be produced at short notice. Unfortunately, there are some researchers who appear all too ready to do this. It does our cause no good at all if we claim too much, and promise more than we are able to deliver.

As Andrew Schonfield wrote, when he was Chairman of the Social Science Research Council in Britain:

> In the social sciences it is rarely possible to pose questions and provide answers in the manner of some of the natural sciences, and it is a refusal to recognise this that has often led us up the wrong path. It is the nature of most of our work that it tends to produce useful ideas and an increasingly firm factual base, rather than clear-cut answers to major policy questions. We must try to tease out the relationships which have a crucial effect on policy and, in doing so, provide not so much widely applicable generalisation as a sound, informed basis for decision-making and, at the same time, cut down the area of reliance on guesswork and prejudice.[11]

It is on these terms that we have to try to be accepted, and this is no easy task.

When we deal with policy research, or policy-oriented research, we must recognise that we are also confronted with other problems about the nature of social science. There are those who insist that, in the final analysis, social science should never accept an exclusively therapeutic or problem-solving role. If both the aims and instruments of research are controlled, as they could be, how can there be the autonomy and independence of enquiry that some would claim is the *sine qua non* of any truly scientific endeavour? When we make our research recommendations, plan our strategies of intervention, and seek greater involvement, can we avoid the clash between policy interests on the one hand and the requirements of social scientific enquiry on the other? Or even more fundamentally, are we in fact in agreement as to what the basic requirements of social scientific enquiry really are? Irving Horowitz argued, many years ago, that where policy needs rule the critical effort would be the exception rather than the rule, and deterioration in the quality of social science would be inevitable.[12] Are we sufficiently aware of this danger when we make our proposals?

In fact, Horowitz maintained that the realities of the situation were such that the utility of the social sciences to policy-making bodies depended on the maintenance of some degree of separation between policy-making and social science. Following this line, some years ago I made a distinction between policy research and policy-oriented research.[13] The former I saw as simply serving the policy-makers on their terms; the latter as addressing the same policy issues (or at least including such issues on the research agenda), but addressing them externally and independently, and with a view, where appropriate, to proposing alternatives with regard to both means and ends.

Relationships between social science and policy will obviously vary from country to country. In some countries the research effort is geared entirely to national policy, and there is a clear understanding of the role of the social scientist. In others, the two spheres might be formally regarded as completely independent of each other, but in practice different parts of the research sphere will probably have different relationships with the policy sphere. In principle, the pattern can vary from complete servitude to genuine critical independence, but there is more than a suspicion that independence and purity are usually inversely related to power, status and influence in decision-making. In this sort of situation there is almost bound to be considerable confusion and uncertainty about the role of social science with regard to policy.

Let us take our line from Horowitz who believes that our main aim, as researchers or social scientists, is to contribute to making society a better place to live in.[14] He maintains that we can do this by transcending rather than by accepting the square world of political and sociological consensus. We do not have to be over-concerned with the restitution of normative patterns, nor need we fall into the trap of examining the costs of dissensus and ignoring the price we pay for consensus. We should address ourselves to social problems without necessarily identifying ourselves with the values of the establishment.

There could of course be another obstacle to our progress, involvement, acceptance and penetration, and this has to do with the very nature of social science, as well as the way in which the researchers, their outlook, their work and its application are regarded by others, such as policy-makers.

Let us assume that when we make our proposals we are asking to be taken seriously. We are claiming that we have a worthwhile contribution to make. But this might be questioned. How good is our past record? What have we contributed? It has been said that when we are not trivial we are contentious and dogmatic, and that we are rarely relevant. Whatever the truth of this, surely we are not justified in attributing all of our shortcomings in this connection to the external obstacles, opposition and hostility referred to earlier. Perhaps we shall have to put our own house in order before others will take us seriously.

We have to consider the possibility that, to the outsider, including the policy-maker and the media practitioner, we may not present a very convincing picture. This is not simply because the general field is inhabited by scholars from different disciplines, with different values, aims and purposes, who seek to construct reality in their own ways. Pluralism in this arena is as necessary as it is now seen to be in other spheres. The complexity of our subject matter and the embryonic stage of development of our subject are among the factors that make this inevitable. Complementary perspectives are essential.

The main problem is akin to the one described by John Rex some years ago.[15] In commenting on 'the crisis in sociology', Rex wrote of warring schools which were characterised more by dogma, doctrinaire assertions, selectivity, arrogance and intolerance than by the respect for evidence, careful examination and description, caution and the consideration of alternatives that ideally one might expect from social science.

That validation and disciplined, systematic study should be given priority over assertion does not imply indifference to values and social concerns, nor should it prevent us from advocating and working towards preferred futures and having our own specific aims and objectives. We do need to recognise, however, that others may have different preferences and objectives. In fact, it is the commitment, the social concern and the wish to use results to produce change that gives research not only its dynamic quality, but also its justification. As Alvin Gouldner argued several years ago, the critical, moral component is a vital part of an endeavour which is essentially purposive, and in which social scientists might be likened to 'clinicians striving to further democratic potentialities'.[16]

THE NEEDS OF THE THIRD WORLD: RESEARCH IMPERIALISM?

The important thing, whatever standpoint or framework of interpretation is adopted, is that prescriptions for the future and recommendations for change should be based on systematic and accurate observations, descriptions and analyses of past and present reality. The research programme that most of us would like to see developed and accepted cannot possibly enjoy the security and safety of the ideological package tour, where every decision and all arrangements have been made beforehand. There are researchers in our field who appear either unable or unwilling to make the distinction between ideology and social science, and often promote the former in the shape of the latter. If we fail to recognise this then we do not deserve to be accepted and we are not likely to have the impact we seek.

The problems already mentioned are exacerbated in a Third World context, where we are likely to experience more discontinuity and dissonance and even less consensus than in our own societies. It has been suggested that a research imperialism has been added to the other imperialisms with which we have become acquainted. But, as this further problem might also stem, at least in part, from the previously mentioned failure adequately to understand the nature and scope of our own research activities, we now need to look at some of the possible implications of these basic problems, these differences, and this lack of consensus.

The situation to which I have briefly referred, with its discontinuities and lack of consensus, becomes even more problematic when we add geographical and stage-of-development components to the differences already mentioned. One has only to have experience of an international comparative research exercise to realise that cultural, regional and national differences profoundly influence the research effort at all stages and levels.

In the circumstances it is inevitable that we have disagreements as to aims, purposes, needs, theories, conceptualisation, design, methods, the nature of evidence and validity. In such circumstances dialogue, meaningful exchanges and constructive debate become extremely difficult, for the necessary agreed referents and commonly accepted basic assumptions are not present. Nor is this just a problem associated with the embryonic stage in our development, although there are those who feel we shall eventually rid ourselves of the conflicts, contradictions and discontinuities and reach the holy grail of consensus, the hallmark of 'real science'.

But realistically, given the essential and inevitable contestability of the theories and methods of social science, what are the chances of this? Is it an objective which we should pursue? If consensus is a mark of scientific maturity, as some would claim, then the social sciences are not very mature. But what is more to the point is that it is possible that, by their very nature, the social sciences will never be able to grow up or mature in this way. This could mean that new criteria of development and maturity are required, such as 'healthy disciplined dissent'.

In suggesting this I am by no means wishing to signal the end of our research effort. Systematic, disciplined, fruitful studies can still be carried out within an eclectic framework, and assessed accordingly. This is not an escape from rigour, but an acceptance of an approach (albeit as yet by no means a fully developed approach) which is capable of doing justice to the complex set of relationships, structures and processes which characterise our field of study.

We must now ask how does this condition relate to research in or about Third World countries? What are we exporting from the so-called developed world? How suitable are these exported models for the conditions it is intended that they should address? Are political, commercial, cultural and media imperialisms being followed by a research imperialism? What forms of indigenisation (or 'native developments') are required, and to what degree should they be applied? These are just a few of the questions which should be asked, both directly in relation to our research, and more generally and more widely with regard to universality and relativity in the social sciences.

When we examine social science research within the international context, and take into account exports and imports of textbooks, articles and journals; citations, references and footnotes; employment of experts (even in international agencies); and the funding, planning and execution of research, then it becomes clear that we have yet another example of a dependency situation. This is a situation which tends to be characterised by a one-way flow of values, ideas, models, methods and resources from North to South. It may even be seen more specifically as a flow from the Anglo-Saxon language community to the rest of the world, and perhaps even more specifically still, within the aforementioned parameters, as an instance of a one-way traffic system which enabled US-dominated social science of the conventional nature to penetrate cultures in many parts of the world which were quite different from the culture in the US. It has been argued that, as the US emerged as a super-power in social science, like it did in other spheres, even what little input was available from other sources tended to be excluded.

It is widely recognised that, in terms of communication research (interests, theories, concepts, methods and findings) much of what was exported from the US (post-World War II) and the implications of these exports were on the whole detrimental. The exports certainly did not serve to increase our understanding of the Third World and its communication requirements, nor did they facilitate development.

Daniel Lerner's extremely influential work *The Passing of Traditional Society* was a prime example of this, irrespective of whether or not it is regarded as an artefact of the Cold-War politics of that time.[17] However, it has been argued that this is not simply a matter of unsuitable exports – it is a much more fundamental matter of bad social science *per se*. The point being made here is that the principles and models underpinning this type of research would not have been adequate in *any* situation, including the situation in the USA. To export such models simply compounded the felony, so to speak. It was not solely a Third World problem, although it certainly was this – it was essentially a social science problem.

This takes us back to the questions already raised about the very nature, potential and universal applicability of social science, no matter how free it may be from the aforementioned conditioning. We have plenty of basic problems at the national or regional levels, but we must now ask how can we possibly deal with the increasing diversification within our general field of communication research which inevitably stems from the extension of our investigations to cultures outside the cultures within which most of the ideas and tools were conceived, developed and articulated?

In general terms, the answer frequently given to this question is 'Indigenisation at Several Levels'. Unfortunately, this proposed solution is often put forward without any apparent recognition that, in certain circumstances, it could lead to increasing dissonance, discontinuity and lack of consensus.

The cry for the indigenisation of social scientific and mass communication research cannot be dismissed, but it needs to be treated with reserve in certain areas, particularly in relation to some of the ways in which it has already been applied. We may readily accept the need for emerging nations and regions to determine their own research policies, priorities and strategies, rather than having them externally imposed, as was the case so often in the past, and the need for home-based institutions, housing native staff capable of carrying out the necessary research in their own countries, also appears to be generally acceptable – at least on the surface. I insert this 'surface' qualification simply because, for many years now, the case has been fiercely argued that the situation would improve to the benefit of Third World countries if only the nationals of those countries could be given the opportunity, and the resources, to enable them to carry out the research. But this is far too simplistic a view, as our experience makes clear, for many of these nationals have been trained as conventional researchers, mostly in the West, and seem unable – perhaps sometimes unwilling – to free themselves from the ideological shackles of their educational and professional mentors. In this way they may even exacerbate the situation and perpetuate the error by giving the 'alien import' a national seal of approval.

The heart of the problem is at the level of language, conceptualisation, models, paradigms, theories and methods. The task here (assuming that the aim is still to pursue a form of universalism rather than encourage parochialism or complete relativism) is to identify, recognise and accept emerging indigenous phenomena in their own right, and then attempt to integrate them into a more universalistic framework. This would be done sympathetically and systematically, but with a full recognition of the inevitability of eclecticism in the foreseeable future. As Habermas and others have suggested, the problem

> cannot be overcome by confrontation or by a nostalgic retreat into old ways. Rather the solution must lie in a recourse to reason and joint reflexion, with a view to developing new norms better adapted to our times, and in a move towards a universality, which would take into account the diversity of cultural identities. The problems of universality and relativity of values, of individual liberty and life in community, of autonomy and solidarity cannot be avoided, and must be resolved.

This, surely, is the democratic task which we have to attempt.

GLOBALISATION, ACCOUNTABILITY, MORAL RESPONSIBILITY AND HUMAN RIGHTS

One can hardly write about Developments in Communication and Democracy without making some reference to the changes in Eastern Europe in recent years. Much has been claimed about the role of the media in this connection, but as yet hard evidence from research in support of the claims is in short supply.

Eric Hobsbawm, reviewing the situation in Europe, and paying particular attention to the tensions between globalisation and nationalism, suggests that when the old established certainties and psychological anchors have gone, new ones have to be found. In such circumstances, either singly or in combination, nationalism, ethnicity, religion, separatism and various forms of fundamentalism offer safe fall-back positions; 'xenophobia looks like [it is] becoming the mass ideology of the twentieth century *fin de siècle*'.[18]

This need for certainty and a sense of identity is not confined to the old Eastern bloc. It is also present in capitalist societies. Two strands may be identified in this connection. The first has to do with the removal of the unifying, integrating functions of the Cold War, which has led to a turning inwards and to a search for and possible identification of 'the enemy' nearer home. The second is the uncertainty and lack of identity which stems from the alienation of the dispossessed as the bonds and networks that provided a sort of social cement have been progressively eroded.

John Kenneth Galbraith believes that, in certain countries, there has been the breaking of a social contract which held that while some might be richer and some poorer, all could expect both protection and a chance in the social system.[19] In the 1980s this understanding was, to put it mildly, placed in abeyance. Kevin Phillips writes of a decade of greed, self-indulgence, rising debt and false prosperity, yet with no real strength or self-confidence – a decade where the gap between rich and poor, both within and between nations, increased considerably.[20]

Other critics remind us of a largely black or immigrant inner-city underclass of single parents, poorly educated people without jobs, opportunities, family values or hope.[21] This raises the possibility of disintegration as the centre collapses and the ideal of the melting-pot gives way to a splintering into racial and ethnic groups which are more concerned with group identity and interests than with integration.

But, as well as being the decade of deterioration, the eighties was also the decade of deregulation, privatisation, increased competition, supremacy of market forces and the rolling back of the public sector. All of these tendencies are energetically pleaded by the advocates of globalisation and technological expansion. They speak primarily (at times solely) in terms of the benefits (unspecified) that 'will inevitably stem' from technological developments. The possible wider social, cultural and political implications are ignored, as are the possible relationships between the two aforementioned sets of tendencies.

Galbraith believes that if this move to the market was the only remedy for socialist countries, then the only way out for Western industrial societies is 'a large, more effective compassionate role for the State'. But this compassionate role is not much in evidence. The last decade has witnessed a progressive erosion of the public sphere of activity. As Reagan saw it, 'Government is not the solution to problems, but *the problem*' and, according to Lady Thatcher, 'There is no such thing as society'. It remains to be seen if such attitudes will persist.

Galbraith holds that western conservatives became so euphoric in the wake of the happenings in Eastern Europe that this blinkered them to the real nature of their own condition. Consequently, they are in grave danger of becoming incapable of seeing the faults in their own system. The eighties might have been the decade when dogmatic socialism went out of fashion, but it now seems possible that the dogma of unfettered free enterprise is also beginning to find itself in trouble. Those who escaped from authoritarian regimes are already beginning to realise that all is not green on the other side of the hill.

Turning to the Third World – a target for the advocates of globalisation – we must bear in mind that the technological developments, in communications and elsewhere, which it was claimed would improve the quality of life do not take place in a vacuum. They are conceived, planned, funded, organised, facilitated, operated and exported within a deregulated free market system. It is this system which 'gives globalisation its dynamic'. It operates at several levels and consists of interrelated parts, of which technological development in communications is but one. But the parts are not exported and do not travel in watertight compartments. They have their value components which are an integral part of a larger economic/social-cultural package which both reflects and reinforces the total system.

The ideologies and conditions obtained in powerful countries have implications far beyond the shores of those countries. The Houston summit of a few years ago, in dealing with farm surpluses, subsidies and dumping, provided a powerful example of how the free market doctrine leaves the developing world still hungry. Experience indicates that globalisation, through the operations of transnational companies, is frequently inappropriate to national needs in Third World countries, primarily because market-place criteria prevail. Recently, the Minister of Health in Brazil, speaking of the unbalanced nature of public development, referred to an alienation linked to priorities that were 'more material than social, more private than public'. A well-researched study in Brazil, *Cultural Genocide*, in illustrating the primacy of consumerism, shows how radio which began as a channel for culture was taken over for economic gain. The conclusion of this detailed work is that the media make use of a public good (the electromagnetic spectrum) in order to serve the economic interests of elite groups – ignoring the needs and privations of the poor.[22]

Noam Chomsky widens the debate about the role of the media from agents of international consumerism to the manufacturers of consent.[23] The media are seen as subtly mobilising public support for special elitist interests, homogenising a narrow spectrum of 'thinkable thought', and creating tacit approval for exploitation.

But there are many countries far worse off than Brazil, and whatever figures one wishes to use (GDP, trade, industrial capacity, debt, use of energy, education, health, domestic savings, military spending, etc.) the imbalance and inequalities between the 25 per cent of the world population who have and the 75 per cent who do not have are quite astounding. For example, the 'haves' consume 80 per cent of world energy, have 86 per cent of world industry (five countries control 60 per cent of industry, and the 44 least developed countries have 0.21 per cent). On the other hand, the rich pay only four per cent interest on debt compared with 17 per cent paid by the poorer nations. The net annual out-flow from South to North in debt repayments amounts to $50 billion (US). Moreover, despite promises from the wealthier countries, the situation is not improving. Forty countries are worse off now than they were a decade ago. Globalisation, as currently conceived, seems likely to exacerbate this. The media have little to say about these conditions.

These problems were high on the agenda of the Earth Summit held in Rio in 1992, and once again the world leaders promised to help the developing world. But the chances of the promises being kept are very remote. Debt continues to drain the 'have-nots' – sub-Saharan Africa in particular. As far as trade is concerned, the protracted dispute between the EEC and USA on the GATT is also to the detriment of Third World interests. Currently, northern protectionism costs the 'have-nots' more than they receive in aid. An examination of the international scene provides a depressing picture of a global class system in which the sands of gross poverty are carried on the winds of change and development.[24]

This, then, is the inescapable arena of globalisation and technological developments in communication. Yet the globalisation experienced so far through the operations of the transnational corporations, which are so uncritically and enthusiastically espoused by the media, seems unable and/or unwilling to meet Third World needs. This primarily is because of the prevalence of market-place criteria and the absence of public concern and accountability. As it currently exists, globalisation implies continued dependency. I need hardly mention that what research is carried out in this general area barely addresses the social, cultural or moral implications of globalisation.

What we really and urgently require is a globalisation of moral responsibility. In the first instance this might enable us to diagnose the problem correctly by carrying out critical research at the appropriate time. This is an essential first step – the prerequisite of the equally necessary education and action which could then follow.

The situation demands public concern, public involvement and public accountability, although to adopt this position is not to advocate one extreme to counter another. The practical solution might be a socially pragmatic combination of private and public activity. However, those who do not believe in the benefits of unfettered free enterprise and wish to explore further the concepts of responsibility, accountability and democracy might look at Dahrendorf's concept of citizenship – 'a system of right and entitlements which will embrace the whole of society'.[25]

Dahrendorf outlines three sets of basic rights: those concerning justice and equality (normally associated with the rule of law); basic political rights (including voting and freedom of expression); and elementary rights (including 'the right not to fall below a certain level of income, and the right to education'). Obviously these elementary rights include vital information and communication components – the *sine qua non* of effective functioning in the information society. But these rights, which are universal, are rarely met. Inequalities mark most societies and there is no trickle down effect at either national or international levels. In fact, we have marginalised underclasses and this itself undermines the very principle of citizenship.

Dahrendorf recognises the illogicality of citizenship being confined to the nation state. He seeks a wider framework of expression which would make it possible to deal more equitably with the foreigners, aliens and migrants in our midst. He also is fully aware of the danger of 'returning to the tribe'. Dahrendorf's main problem is with 'fundamentalism', which he sees as stemming from the anomic conditions of modern life. He puts great store on the exercise of the right to be different, and regards as dangerous those proliferating groups which value homogeneity above the rights of citizens. This at least provides us with some principles to discuss in relation to our theme of developments in communication and democracy.

As the information society develops it will not be possible to achieve the goals of citizenship, or to exercise the appropriate rights and responsibilities in the absence of information and communication systems that provide the information base and the opportunities for access and participation for *all* citizens. Accountability and responsibility demand that those who espouse development and globalisation take this into account. We must also realise that, if we wish to alleviate the conditions of the many 'have-nots' – particularly in the Third World – then some form of self-sacrifice on the part of the 'haves' is essential. The acceptance of this form of moral responsibility with all that it implies in terms of research and policy is central to our concerns.

Unfortunately, the way things are at this time is not particularly encouraging. The environment is generally hostile to what is proposed here and, at times, we as researchers may even be our own worst enemies. Even so, this closing chapter should not be seen as a message of pessimism or despair. Essentially it is a plea for self-reflection, for self-criticism and for a thoroughgoing and honest examination of our strengths and weaknesses. Without this realistic evaluation of our performance and potential we shall never ask the right questions, and consequently we shall never be in a position to direct our efforts to the attainment of our declared goals. Admittedly, given the opposition, asking the right questions is no guarantee of success, but it is the first essential step – a *sine qua non* of effective action.

Notes

1. UNESCO, *Proposals for an International Programme of Communication Research* (COM/MD 20, Paris, 1971).
2. James D. Halloran, *Mass Media and Society: The Challenge of Research* (Leicester: Leicester University Press, 1974).
3. E.J. Mesthene, 'Prolegomena to the Study of the Social Implications of Technology' in R.P. Morgan (ed.), *The New Communication Technology and its Social Implications* (London: International Broadcast Institute, 1971).
4. *Journal of Communication*, Vol.39, No.3 (Summer 1989); *InterMedia*, Vol.18, No.3 (June/July 1990); *Critical Connection* (Congress of the United States Office of Technology Assessment, January 1990).
5. *The Myth of the Information Revolution* (Beverly Hills: Sage Publications), 1986.
6. William H. Melody, 'The Information in I.T.: Where Lies the Public Interest?' in *InterMedia*, Vol.18, No.3 (June/July 1990) pp.10–18.
7. James D. Halloran, *Mass Media in Society: The Need of Research; Unesco Reports and Papers on Mass Communication*, No. 59 (1970); and 'Research in Forbidden Territory' in G. Gerbnes, S.P. Gross and W.H. Melody (eds), *Communication Technology and Social Policy: Understanding the New Cultural Revolution* (London: John Wiley & Sons, 1973) pp.547–53.
8. A. Gillespie and K. Robins, 'Geographical Inequalities; The Spatial Bias of the New Communications Technologies' in *Journal of Communication*, Vol.39, No.3 (Summer 1989) pp.7–18.
9. Melody, 'The Information in I.T.'.
10. The proceedings were published in 1973 under the title *Communication Technology and Social Policy: Understanding The New Cultural Revolution*, G. Gerbner, S.P. Gross and W.H. Melody (eds).
11. A. Schonfield, 'Introduction to the Annual Report' in *Social Science Research Council Newsletter Special* (London: SSRC, 1971).
12. Irving Louis Horowitz, *Professing Sociology* (Chicago: Aldine Publishing Co., 1968).
13. James D. Halloran, 'Mass Communication: Symptom or Cause of Violence?' in G.C. Wilhoit and H. De-Bloch (eds), *Mass Communication Review Yearbook*, Vol.1 (Beverly Hills: Sage Publications 1980) pp.432–49.
14. I.L. Horowitz, *Professing Sociology*.
15. John Rex, 'British Sociology's Wars of Religion' in *New Society* (11 May 1978) pp.295–7.

16. A.W. Gouldner, 'Metaphysical Pathos and the Theory of Bureaucracy' in *American Political Science Review*, Vol.49, No.2 (June 1955) pp.506–7.
17. Daniel Lerner, *The Passing of Traditional Society* (Glencoe, Ill.: Free Press, 1958).
18. Eric J. Hobsbawm, 'Nationalism and Ethnicity' in *InterMedia*, Vol.20, Nos. 4 & 5 (August/September 1992) pp.13-15.
19. John Kenneth Galbraith, 'Assault: An Overview' in *Weekend Guardian* No.16/17 (December 1989) pp.15-17, and Galbraith, 'The Price of World Peace' in *The Guardian* (8 September 1990) p.23.
20. Kevin P. Phillips, *The Politics of Rich and Poor* (New York: Random House, 1990).
21. See, in particular, Noam Chomsky, *Deterring Democracy* (London: Verso, 1991).
22. O.S. Oliveira, *Genocido Cultural* (Sao Paulo: Edicoes Paulinas, 1991).
23. Noam Chomsky, *Deterring Democracy*.
24. Paul Harrison, *Inside the Third World* (London: Penguin, 1993).
25. Rolf Dahrendorf, *Reflections on the Revolution in Europe* (London: Chatto & Windus, 1990).

Index

Printed in the United States
By Bookmasters